Waste to Bricks: Eco-Alternative

Mathew

TABLE OF CONTENTS

Chapter 1 Introduction

This study investigates the use of a natural process called microbial induced calcium carbonate precipitation (MICP) to 'grow' bio-bricks using the urea present in human urine. Furthermore, the study develops a human urine bio-brick production methodology and assesses the various factors affecting the optimisation through review and laboratory experimentation. The introduction discusses the background and motivation, the research gaps, the research aims and objectives for the proposed project and finally, it gives a brief outline of the structure of the book.

1.1 Background and motivation

The global human population was estimated at 7.6 billion in 2017 and continues to grow exponentially; it is predicted to reach 9.8 billion by 2050 (Nations, 2017). Accompanying this population growth, emerging economies are expected to increase investment in the building of new infrastructure. While developing countries are challenged with the task of remediating and maintaining deteriorating infrastructure built in the last 100 years. This predicted development will inevitably result in an increase in the consumption of building materials (Schneider et al., 2011). The production of cement alone accounts for 6% of global CO_2 emissions (Achal & Mukherjee, 2015), putting building materials among the most significant contributors to anthropogenic greenhouse gas emissions. The unprecedented rate of climate change and the associated pressures, coupled with the increased awareness around the depletion of natural resources, presents a significant challenge for which innovative and sustainable solutions are essential.

Altering our building practices offers an opportunity to reimagine the construction of our future cities. A potential starting point is achieving sustainability through the transformation of traditional building materials, initiating a shift toward selecting building materials that are produced by sustainable practices and ensure smaller life-cycle costs. For millions of years, nature has been sustainably producing materials with properties comparable to present day building materials through the process of bio-mineralisation (Achal & Mukherjee, 2015). The work presented in this book focuses on the application of bio-mineralisation processes to manufacture building materials, specifically masonry bio-bricks.

Bio-mineralisation is an emerging process that has been successfully demonstrated in field applications and laboratory work (Achal & Mukherjee, 2015). The core incentive of this work is to produce bio-based bricks as an alternative to the standard brick, which is traditionally produced from clay and fired in a kiln at high temperatures (Zhang, 2013). In 2013, the annual global production of bricks was approximately 1391 billion units (Zhang, 2013) and this production rate is only expected to increase. In addition, quarrying clay generates large amounts of waste and is also an energy-intensive process that can be detrimental to the surrounding environment. Clay is a finite resource, and many countries are expected to experience shortages in the near future (Zhang, 2013). Excessive amounts of energy are required to maintain the high temperatures specific to brick-making kilns, contributing substantially to global greenhouse gas emissions. One unit of conventional clay brick releases 0.41 kg of CO_2 and utilizes 2.0 kWh of energy in its production (Zhang, 2013). Conversely, microbial precipitation is a natural process that uses various biological processes at ambient temperature to produce calcium carbonate. The calcium carbonate cements loose sand particles together. This process is not energy intensive, and the bio-solids can be any shape.

A company in the United States of America (USA), BioMASON, is already using Microbially Induced Calcium carbonate Precipitation (MICP) to grow bio-bricks. BioMASON has scaled its process to reach a manufacturing capacity that can inevitably divert and absorb market demand for a more sustainable brick solution. However, in their process, BioMASON uses synthetic urea produced by the Haber-Bosch process. The Haber-Bosch process uses about 1% of the global energy demand (Chesworth, 2008) and has a low yield, which could counteract the clean and sustainable methods used to produce BioMASON's bio-bricks. Additionally, Previous practitioners have produced mud bio-bricks using human urine (Magnusson, 2011), and other researchers have used pig urine as an alternative source of urea for MICP (Chen et al., 2018). Bio-bricks have also been produced using MICP and synthetic urea (Bu et al., 2018) but no researchers have investigated using human urine in MICP for the manufacturing of bio-bricks. This work, therefore, outlines the experimental procedure for producing the world's first bio-bricks using human urine as a source of urea in the MICP process.

The current linear flow of materials and energy between nature and the economy exacerbates the depletion of ecosystems and natural resources, while simultaneously producing emissions and waste. Applying a circular economy is crucial to reaching a more sustainable future, where the linear process is reversed, and the circular flow of materials is implemented (Korhonen, Honokasalo & Seppala, 2018). Urine is an example of a "waste" product that can be incorporated into a circular flow of production. It contains large concentrations of urea (Rose et al. 2015) and has the potential to replace the synthetic urea required for MICP processes. If MICP were to use urea sourced from urine, it would offer a viable replacement to conventional masonry units – drastically reducing the water and energy required in the manufacturing of conventional bricks. Conversely, the urea in urine degrades into carbonate ions and ammonia within 24 hours by a process called enzymatic urea hydrolysis (Udert et al., 2003). The carbonate ions produced during enzymatic urea hydrolysis can be used in the production of calcium carbonate. However, the amount precipitated might not be enough to cement loose material together, like in the MICP process. Instead, MICP uses bacteria that act as seeds for the precipitation of calcium carbonate (Anbu et al., 2016), while also creating localised concentration gradients as they degrade the urea. This facilitates improved calcium carbonate precipitation and is a critical step in MICP processes.

In order to use urine in MICP processes, it is necessary to delay enzymatic urea hydrolysis. The inhibition of enzymatic urea hydrolysis can be achieved either by reducing the pH of fresh urine (Ek et al., 2006), by adding urease inhibitors (Kafarski & Talma, 2018) or by increasing the pH through the addition of calcium hydroxide (Randall et al., 2016). The addition of an acid to reduce the pH will also prevent precipitation from occurring; thus nutrients such as phosphorus would precipitate during MICP instead of recovering phosphorus as a solid by-product which can be used as a fertiliser (Meyer et al., 2018, Randall & Naidoo, 2018). The same issue would arise with urease inhibitors. Fortunately, the addition of calcium hydroxide to urine results in magnesium and phosphorus precipitating out almost entirely as a solid fertiliser – calcium phosphate – before MICP (Flanagan & Randall, 2018, Randall et al., 2016). In addition, stabilising the urine with calcium hydroxide means the treated urine can be stored for extended periods of time with minimal urea loss (Flanagan & Randall, 2018).

The challenge faced with adding calcium hydroxide to urine for MICP is the resultant high pH of the solution. For example, urine stabilised with calcium hydroxide would have an approximate pH of 12.5 at 25°C (Randall et al., 2016) in which urease-producing bacteria do not survive. The urease produced by these bacteria is a catalyst for the degradation of urea (Mobley & Hausinger, 1989) and essential for any MICP process. The pH threshold for this to occur was found to be 11 (Randall et al., 2016) and Henze & Randall (2018) showed that *Sporosarcina pasteurii* (*S. pasteurii* bacteria (urease producing bacteria commonly used in MICP studies) could survive at a pH of 11.2. In fact, the rate of MICP was improved at these elevated pH values because more carbonate ions are present at elevated pH values. These researchers successfully grew bio-columns using MICP and synthetic urine at the increased pH values (Henze & Randall, 2018).

The production of nitrogen gas for fertilisers by the Haber-Bosch process and the production of phosphorous fertilisers from the mining of phosphate rock are exemplary of the linear usage of materials and energy. Phosphate rock reserves are a finite resource and only found in certain countries (Cordell & White, 2014). It has been predicted by some researchers that phosphorous reserves will hit their peak in 50 years (Cordell & White, 2014). A portion of the nitrogen and phosphorous fertilisers used are released into receiving bodies of water as run-off, along with

(Rittmann et al., 2011). Additionally, about 16% of all the mined phosphorous will end up at conventional wastewater treatment plants, of which 8% is released into receiving water systems (Rittmann et al., 2011).

The function of a wastewater treatment plant is to safeguard the receiving water bodies against eutrophication and subsequent degradation from organic and nutrient (nitrogen and phosphorous) pollution. The denitrification process and precipitation of phosphorous out of wastewater are both energy intensive processes using between 45 and 50 MJ per kg of nutrient (Maurer & Schwegler, 2003). Often these processes are inefficient and cannot cope with the influent load. During denitrification, nitrogen, a valuable resource, can be lost in its gaseous form. In addition, nitrogen together with phosphorous can be lost as sludge and incinerated, ending up in the effluent or a landfill. Sewage treatment plants release about 7.4 million tonnes of nitrogen (Van Drecht et al., 2009) and 1.5 million tonnes of phosphorous (Rittmann et al., 2011) annually, subjecting water bodies to further nutrient overloads. The possibility of recovering these nutrients from wastewater could prevent the depletion of mineral deposits and reduce the amount of energy utilised to make them available.

Alternatively, human faeces and urine are rich in nitrogen and phosphorous, and domestic wastewater has the potential to provide 22% of our global phosphorous demand (Mihelcic, Fry & Shaw, 2011). Source separation of urine from other domestic wastewater components can facilitate the recovery of nutrients. The urine stream contains 50% of the phosphorus and 75% of the nitrogen (Larsen & Gujer, 1996) in domestic wastewater. The benefit of separating urine is that both phosphorous and nitrogen could be recovered as inorganic fertilisers on site. The recovery of nutrients directly from human urine has the potential to be used in agriculture while preventing pollution and recovering financial value through the sale of recovered nutrients as fertilisers (Randall et al., 2016).

1.2 Problem Statement

Seeking alternative methods for producing masonry products and recovering nutrients from human waste is pertinent for building a sustainable future. Many benefits can potentially be achieved by producing bio-bricks using human urine as a source of urea for MICP. Methodologies for producing masonry products from MICP do exist but none use urine as a source of urea.

1.3 Objectives & Scope

The overall aim of this work is to test the feasibility of using urea, sourced from human urine, to produce bio-bricks using MICP and subsequently optimise the process. The project has the following key objectives:

1. Conduct an in-depth literature review on the theory of MICP, the factors affecting MICP, previous research and industry application of MICP and finally urea hydrolysis and urine stabilisation methods.
2. Test the possibility of producing a bio-brick from human urine using the MICP process.
3. Improve the process design to determine the optimal influent calcium concentration, retention time and the number of treatments in relation to the resultant compressive strength.
4. Establish a relationship between treatment duration and bio-brick strength.
5. Understand how the ionic strength and the pH of the urine being fed into the bio-brick mould, affects the MICP process.
6. Reduce the cost of the process by investigating industrial waste streams as a viable, alternative food source for the bacteria governing MICP.
7. Conduct a preliminary economic analysis of an integrated urine treatment system and discuss the social and policy barriers likely to arise if such a system is implemented.

This book expands on research conducted by Henze & Randall (2018) who used synthetic urine to produce bio-solids and designed some of the testing techniques used in this work. To develop this research, urea from real human urine will be used instead of synthetic urine, and the methodology will be implemented to produce a bio-solid in the shape of a conventional brick. Furthermore, this book aims to explore the effect of experimental conditions on the calcium and urea usage efficiency, the time required to produce a bio-brick and its resultant compressive strength.

In addition, the mechanisms governing the microbial interaction needs to be identified and their effective relationships investigated. Such mechanisms include the effect of the ionic strength, calcium concentration and the pH of the influent solution on the MICP process, the calcium carbonate precipitated, the ureolytic activity and the effluent pH.

The 5th objective is to reduce the cost of bio-brick production by using an alternative growth media. At present, the process would not be profitable because of the American Type Culture Collection (ATTC® 1376) growth media used for growing the bacteria. A number of experiments need to be conducted to assess the viability of using various industrial waste streams as an alternative growth media. Their viability is assessed by producing growth curves and comparing them to the conventional growth media used in previous studies. The viability of the waste streams is also measured by assessing whether ureolytic activity occurs in the bacteria grown in the specific waste stream, simulating the bio-brick conditions to see if MICP occurs and comparing rates of calcium carbonate precipitation.

The 6th objective includes drawing up a basic process flow diagram for an integrated system and conducting an economic analysis of the operating costs of such a system. The integrated system includes the production of the bio-brick as well as the recovery of inorganic fertilisers before and after the bio-brick production.

Sporosarcina Pasteurii (S. pasteurii) is the specific urease-producing bacteria that will be used in the study as it is non-pathogenic; its optimal growth exists in the alkaline region, and it's commonly used in the application of MICP (Anbu et al., 2016). The identification of nutrient medias referenced in literature, further limited to locally sourced waste streams in the Western Cape region of South Africa and who's company owners were willing to offer samples for testing, determined the various alternative nutrient medias chosen for testing. The economic analysis conducted in the study is based on the integrated system being implemented in a local, South African context.

The study collected urine from male subjects only, in a waterless, nutrient recovery urinal, that also stabilised the urine by the addition of calcium hydroxide (Flanagan & Randall, 2018). The frequency at which the urine was collected was subject to student's and staff's willingness to contribute their urine and how often they frequented the bathroom where the urinal was located.

The study focuses on the data collected from repeated experimental runs using specifically designed moulds, informed by research, previous studies and with the equipment available in the labs. The investigation is limited by access to funding and the constrained time frame in which the study must be completed. The data collected together with the relevant literature inform the conclusions drawn for the optimal conditions of the bio-brick production process. The lack of research specifically geared towards the objectives of this book meant that further experiments needed to be conducted; to understand how using stabilised urine as a cementation medium affected the process, with concern to its high pH and the other ions and compounds present in human urine.

1.4 Plan of Development

This book comprises of 8 chapters. Chapter 1 serves to introduce the research by outlining its background, importance, objectives, scope and limitations. Chapter 2 conducts an in-depth literature review on the research performed in the study, with specific reference to apparent the objectives outlined in chapter 1. Chapter 4 presents the results and research findings from the various experiments conducted. Chapter 5 discusses the results and varying interpretations while comparing any deviations or similarities to other research. Chapter 6 addresses the economic and social impact of implementing the research into the industry by conducting a broad economic analysis of the process and discussing the social inhibitors. Chapter 7 serve to conclude and synthesise the empirical findings of the conducted research and make recommendations on further research.

Chapter 2 Literature Review

This chapter conducts an in-depth literature review to understand further the topics touched on in this book and inform decisions for experimentation and analyses. Firstly, research pertaining to the understanding of Microbially Induced Calcium carbonate Precipitation (MICP) is assessed. Subsequently, the factors that affect the mechanism of MICP and urea hydrolysis are specifically examined.

The Literature review further summarises both the current research and field application of MICP existing today. Moreover, research pertaining to how calcium carbonate forms between grains, the crystalline nature of the calcium carbonate forming and the factors that affect both are reviewed. Compressive testing standards for masonry brick products and the bacteria strain to be used in this book were also surveyed.

Additionally, the problems pertaining to urine collection, specifically urea hydrolysis, and the existing methods on how to inhibit it through urine stabilisation are reviewed to inform the urea stabilisation method proposed in this work.   Finally, a brief discussion of research done pertaining to the proposed urine stabilisation method is also given.

2.1   Microbial Induced Calcium Carbonate Precipitation (MICP)

Microbial induced calcium carbonate precipitation (MICP) uses the natural process, bio-mineralisation, to precipitate calcium carbonate. Bio-mineralisation is a phenomenon that uses living bacteria and microbial activity to cause a reaction between metabolic products and their surrounding environment which results in the formation of a mineral. Two different mechanisms exist by which bio-mineralisation can take place (Muynck, Belie & Verstraete, 2010): biologically controlled mineralisation (BCM) and biologically induced mineralisation (BIM).

BCM works intracellularly, where minerals are deposited within the organic matrices or vesicles of the living cells (Achal & Mukherjee, 2015). MICP operates according to the second mechanism, BIM, through which biological activity chemically modifies the environment resulting in supersaturation and precipitation of minerals (Muynck, Belie & Verstraete, 2010).

The rate of precipitation is dependent on a combination of factors that cannot be sufficiently controlled and monitored in the scope of this work. However, understanding how the processes work helps produce optimal conditions for MICP. Induced mineralisation is mainly dependent on environmental conditions (Muynck, Belie & Verstraete, 2010). These conditions are made up of four main factors: the calcium concentration, the dissolved inorganic carbon (DIC) concentration, the pH and the available nucleation sites (Muynck, Belie & Verstraete, 2010).

The formation of calcium carbonate occurs according to equation 2-1. In order to favour the formation of calcium carbonate, the system must be oversaturated with both calcium and carbonate ions. The ion activity product (IAP) must exceed the solubility constant $K_{sp} = 4.8 \times 10^{-9}$ for precipitation to occur. The IAP can be described as being directly proportional to the concentration of calcium and carbonate ions according to equation 2-1.

$$K_{sp} = [Ca^{2+}][CO_3^{2-}] \qquad \text{2-1}$$

The concentration of carbonate ions is related to the given system's DIC concentration and the pH (Muynck, Belie & Verstraete, 2010) and increases with rising alkalinity and DIC. DIC is dependent on specific environmental parameters such as if the system interacts with the atmosphere. One such environmental parameter is the DIC equilibrium with CO_2 in the atmosphere. A shift in the equilibrium causes volatisation or dissolution of CO_2.

Two different metabolic pathways exist where bio-mineralisation with microorganisms can occur. Autotrophic mediated pathways use CO_2 as a source of carbon; calcium carbonate precipitates as the microbes convert atmospheric CO_2 into carbonate ions when calcium ions are present in the environment. Heterotrophic-mediated pathways use the nitrogen or sulphur cycle to precipitate calcium carbonate. Sulphur reducing bacteria (SRB) degrade organic matter in an anoxic environment rich in calcium and sulphate ions. With SO_4^{2-} as an electron acceptor, this process produces bicarbonate ions and hydrogen sulphide according to equation 2-2 The degradation of the organic matter causes a rise in pH which is favourable for calcium carbonate formation (Joshi et al., 2017). Equation 2-3 indicates that the calcium sulphate present in cavities can also provide a source of calcium ions.

$$2CH_2O + SO_4^{2-} \rightarrow H_2S + 2HCO_3^- \qquad \text{2-2}$$

$$CaSO_4 \cdot 2H_2O \rightarrow Ca^{2+} + SO_4^{2-} + 2H_2O \qquad \text{2-3}$$

The precipitation of calcium carbonate by the nitrogen cycle can occur by using three different mechanisms: in an aerobic environment that contains both organic matter and calcium, amino acids undergo ammonification; in an anaerobic environment that contains organic matter, calcium and nitrate, the dissimilatory reduction of nitrate occurs, or in an aerobic environment where urea, calcium and organic matter are present, urea hydrolysis occurs. All microbiological mechanisms result in the generation of ammonia and bicarbonate (Joshi et al., 2017). The pH of the solution rises when ammonia is generated, this shifts the equilibrium to favour the production of the carbonate ions which then bond with the calcium ions present in the solution. The third mechanism governs the process used in this book and is also utilised

2.2 MICP by Urea Hydrolysis

Ureolysis (urea hydrolysis) is the process whereby urea is degraded into ammonia and carbonic acid as shown in equation 2-4. Ureolysis is catalysed by the enzyme urease. Urease is a widespread enzyme produced by bacteria and is also produced in some plants (Lauchnor et al., 2015), but for the purposes of this research work, it is produced by the bacteria, *Sporosarina Pasteurii*. A number of different bacteria can produce this enzyme and can be used for the MICP process. These include *Bacillus sphaericus, Proteus vulgaris, Helicobacter pylori, Sporodarcina ureae, Myxococcus xanthus* and *Proteus mirabilis* (Dosier, 2016).

$$CO(NH_2)_2 + 2H_2O \rightarrow 2NH_3 + H_2CO_3 \qquad \text{2-4}$$

MICP by urea hydrolysis generally combines cells or enzymes with calcium ions and urea. The urease producing cell use the urea as a source of energy to catalyse the reaction mentioned in 2-4 to produce ammonia and carbonic acid. The production of ammonia causes the pH level to change so that calcium and carbonate ions form a precipitate. The aggregate or bacteria cell walls act as nucleation sites, forming calcium carbonate crystals that can act as a biogenic cement in porous media (Lauchnor et al., 2015).

2.3 Kinetics of MICP

As this project operates in a closed system, the gas exchange between DIC and CO_2 is limited to zero. After the elimination of this parameter, the DIC can be increased by biological activity. The kinetics of the rate of precipitation of calcium carbonate has been modelled using equation 2-5 (Tobler et al., 2011).

$$\frac{dCa^{2+}}{dt} = -Ksp(S - 1)^n \qquad \text{2-5}$$

Where;
$$S = \frac{[Ca^{2+}][CO_3^{2-}]}{K_{sp}} \qquad \text{2-6}$$

From equations 2-5 and 2-6, it can be seen that the rate of precipitation is dependent on the IAP and is affected by the activity of the ions. If the concentration of the ions is higher, the value of S (the saturation index) will increase such that the rate of precipitation will also increase. The constants, K_{sp} and n, are both dependent and are functions of the crystal growth mechanism, its surface area, mineral type and the number of crystals (Henze, 2017, Henze & Randall, 2018)

The rate of ureolysis also depends on the bacterial cell density and the available urea

Lauchnor et al. (2015) found that the commonly used first-order kinetic model describing enzyme-catalysed reactions is sufficient at most relevant concentrations. Equation 2-7 below represents the consumption of urea.

$$\frac{dS_{urea}}{dt} = V_{max} \cdot \frac{S_{urea}}{K_{m,urease} + S_{urea}} \qquad \text{2-7}$$

$K_{m,urease}$ is the half-saturation coefficient, and V_{max} is the maximum rate of reaction measured in units of urea concentration per time (Lauchnor et al., 2015).

2.4 The Effect of PH, Calcium Concentration and Ionic Strength

Forming minerals enhances the competitiveness of a bacteria species in its microbial community, as it aids in the detoxification of poisonous inorganic species (Herbaud et al., 1998). Intracellular calcium ions (Ca^{2+}) helps regulate essential processes within the cell (Herbaud et al., 1998). Calcium regulation in bacteria cells is governed by two transport mechanisms, either active or passive. Passive regulation occurs between two antiporters ($Ca^{2+}/2H^{+}$, $Ca^{2+}/2Na^{+}$) and requires an electrochemical gradient (Hammes & Verstraete, 2002, Herbaud et al., 1998). Usually, the calcium influx is caused by the electrochemical gradient from intracellular calcium concentrations being around 103 lower than that of the extracellular gradients. (Hammes & Verstraete, 2002). Active calcium transport uses ATP-dependent (adenosine triphosphate) pumps that transport calcium against an electrochemical gradient at the expense of energy (Desrosiers et al., 1996, Hammes & Verstraete, 2002). This process produces ammonia and carbonate ions and if calcium ions are present, calcium carbonate precipitates (Acuna et al., 2017). The proton expulsion from the bacteria causes its surface to become negatively charged, facilitating the adhesion of Ca^{2+} to act as nucleation sites (Acuna et al., 2017).

Several researchers have reported that calcium precipitation efficiency is negatively affected by high urea and calcium concentrations (Cuthbert et al., 2012). Furthermore, Cuthbert et al. (2012) postulated that this is a result of a build-up of calcium carbonate around the cell, encapsulating it and inhibiting the exchange of urea to ammonia. Conversely, (Cuthbert et al., 2012), Hammes & Verstraete (2002) suggest that the reduction of calcium precipitation efficiency and urea hydrolysis is a result of passive calcium influx. Section A in Figure 2-1 shows the passive calcium influx due to an increase in pH and extracellular calcium. The electrochemical gradient increase leads to a build-up of intracellular calcium and excessive proton expulsion, causing localised acidification. The influx disrupts the intracellular calcium regulation, alkalising the intracellular pH and depleting the proton pool, negatively affecting the cell physiological processes and reducing its ability to hydrolyse urea (Hammes & Verstraete, 2002).

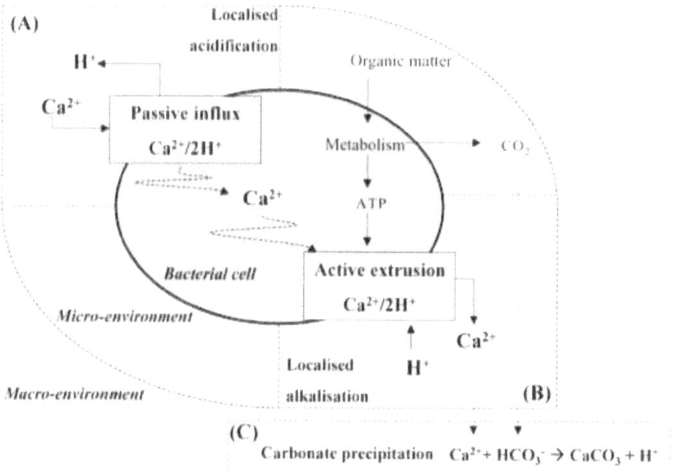

Figure 2-1: A pictorial representation of bacterial calcium metabolism and subsequent CaCO₃ precipitation under high-pH and high Ca²⁺ extracellular condition (Hammes & Verstraete, 2002).

The correlation between these adverse effects has yet to be studied in detail, especially in conjunction with the high pH conditions that are to be used in this work. Literature shows that the pH of a solution increases as ammonia is released during urea hydrolysis to reach the optimum pH for ureolytic activity. When the solution is fully saturated with carbonate ions at the alkaline pH of optimum ureolytic activity, the calcium carbonate precipitates causing a decrease in the pH (Hammes & Verstraete, 2002, Henze & Randall, 2018). To elaborate, calcium carbonate precipitates in an alkaline environment and releases hydrogen ions (H⁺), and as the H⁺ increases, it causes the pH of the solution to decrease. The course of the pH reflects the balance of these two processes.

Howell & Sumner (1934) and Kistiakowsky & Shaw (1953) showed no irreversible denaturation occurring to the enzyme urease in a solution with a pH as high as 10, which means that the application of external stresses of the high pH had not caused proteins or nucleic acids to change their native states. Moreover, experimental data showed that the concentration of hydroxyl ions (higher in an alkaline environment) has no effect on the enzymatic activity, but rather the enhanced ionic atmosphere acts as an inhibitor (Kistiakowsky & Shaw, 1953). Henze & Randall (2018), were among the first to study the effect of a high pH on MICP in a synthetic urine solution using *S. pasteurii*. Their research indicated that the rate of calcium carbonate precipitation was not adversely affected by increasing the pH of a solution to as high as 11.2. It was shown that shifting the higher pH down to around 9 was the optimal pH for urea hydrolysis (Lauchnor et al., urine solution with a small calcium concentration, rather than human urine with an increased calcium concentration. In addition, Henze & Randall (2018) did not monitor pH's effect on ureolytic activity in a quantitate manner measuring it using CUA (Christenson's Urea Agar) plates which only indicated ureolytic activity but did not serve to quantify it. Conversely, experiments by Cuzman et al. (2015) showed that higher alkalinity reduced the rate of hydrolysed urea and suggested that the optimum pH for MICP was 7.

In experiments on the ureolytic activity of urease at pH 8.9, Kistiakowsky & Shaw (1953) found that the activity of the urease was sensitive to a change in ionic strength. Their research showed that the rate of urea hydrolysis was determined by the ionic strength and that the reduction of enzymatic activity is independent of the concentration of the substrate, i.e. urea. Unfortunately, the research failed to describe the mechanisms involved in the effects of the ionic strength. It was postulated that it might be as a result of the urease molecule acting as a zwitterion with the ionic atmosphere surrounding it (Kistiakowsky & Shaw, 1953). A zwitterion, formally named a dipolar ion, has separate positive and negatively charged functional groups of which the net charge of the entire molecule is zero (Hardinger, 2010).

Whilst the basic chemistry of the MICP process is known, research on its more complex interactions is still in its infancy. The current conceptual understanding of MICP lacks information on the *S. pasteurii* bio-film. A bio-film is a layer of an organism which has formed a colony. The colony attaches to a surface with a slime layer which aids in protecting the microorganisms and promotes their growth and survival (Biologydictionary.net, 2017). Research conducted by Hommel et al. (2015) revealed that the concentrations of ammonium and calcium are particularly sensitive to the parameters associated with the behaviour of the bacterial biomass and ureolysis rate. Much of the uncertainty that might be experienced in this book could be understood if a better understanding of the bio-films density, the factors that affect the attachment of *S. pasteuriii* to the grain surfaces or other bio-film surfaces and how this related to the rate of ureolysis (Hommel et al., 2015). The bio-film growth on the grain surfaces can often alter hydraulic flow cutting of highly permeable pathways (Hommel et al., 2015). The cementation media will first fill the more conductive pathways. These pathways with time will fill up decreasing their conductivity allowing the surrounding volume to be filled. This, however, is problematic, as over time the bacterial biomass in the less conductive areas will decay and its cell density will decrease. When cementation media begins to fill the neglected volumes, the decreased cell density may lead to the decreased to ureolytic activity and the bacteria's sensitivity to the pH and ionic strength (Hammes & Verstraete, 2002).

2.5   Applications of MICP

In the building and construction sector, there have been many innovations initiated using bacteria in the process of MICP. MICP has shown to be effective in:

- bio-cementation and crack healing in concrete
- bio-deposition
- bio-remediation
- wastewater treatment (WWT)
- soil strength improvement
- restoration of stone monuments
- increasing durability in concrete structures
- reducing permeability

The subsequent sections discuss these in greater detail.

2.5.1   *Bio-remediation and Crack Healing in Concrete*

Reinforced concrete is one of the most common building materials used today; contributing to approximately 80% of the world's infrastructure. Concrete's popularity is due to its high durability and versatility. It can, however, undergo premature deterioration due to chemical attacks from extreme weather or temperature (Joshi et al., 2017). Remediating and maintaining concrete structures has proven itself to be a great challenge. Early onset formation of micro-cracking in concrete structures adversely affects their serviceability resulting in high maintenance costs (Joshi et al., 2017). Treating the concrete with synthetic agents and applying sealants are the primary methods used to increase the durability of deteriorating structures. These maintenance practices require high amounts of recurring investment and money that many countries cannot afford. In response to this, extensive research has gone into "self-healing" enabled by microbial systems, its application to crack repair in concrete structures and the viability of employing the technology on a commercial scale.

There are three main approaches to self-healing concrete as shown in Figure 2.2 vascular based, capsule-based and intrinsic based. In the vascular-based self-healing mechanism, healing material filled in hollow channels or fibres, embedded in a concrete matrix, is released when damage ruptures the hollow channels or fibres (a). In the capsule-based mechanism, the healing agent stored in the capsules is released when the capsules in the concrete matrix are exposed to external substances and dissolve, for example, water seeping into cracks (b). Intrinsic self-healing materials embedded directly in the concrete matrix possess a latent self-healing functionality that is triggered by damage or by an external stimulus (c). These materials rely on polymerisations, melting of thermoplastic phases or ionic interactions to initiate self- healing

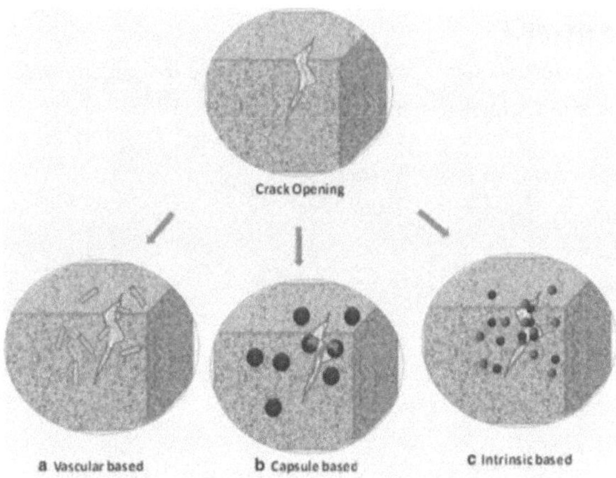

Figure 2-2: Types of engineered self-healing mechanisms within concrete (Joshi et al., 2017).

2.5.2   Bio-deposition and Bio-remediation

The current method of increasing the durability of a deteriorating structure is to make use of synthetic agents, although the durability thereof is often unreliable. Another solution is to employ bio-deposition as a method of remediation. Bio-deposition takes place when a porous surface is exposed to a culture of ureolytic bacteria. Subsequently, MICP takes place, and a layer of carbonate is deposited on the surface. Various methods have been explored to expose the surface to bacteria such as spraying, immersion or ponding (Achal & Mukherjee, 2015). Researchers have found that bio-cementation and bio-remediation have resulted in the reduction of gas permeation and water absorption (Achal & Mukherjee, 2015). Similarly, De Muynck et al. (2008) used the bacterial strain *B. sphaericus* and a calcium mixture to treat the surfaces of mortar cubes in bio-deposition application. The findings showed that the calcium carbonate crystals precipitated on the surface decreased the water absorption by 65-90 % and showed an increased resistance toward freezing and thawing.

2.5.3   Soil Improvement: Bio-grout and Rammed Earth Structures

Grout is a binding agent often used to improve the durability functions of building materials. It may be organic or inorganic material used in the sealing or consolidation of cracks, pores or voids, to improve mechanical properties and permeability of a particular system; whether soil, cementitious or other building materials (Kawasaki & Achal, 2016). Grouting material needs to be able to penetrate intermittent voids, cracks and cavities and therefore needs to be soft, fine and easy to inject in addition to having a high compressive strength and a low shrinkage. Chemical grouting is one of the most popular options, although it is expensive and has adverse

Alternatively, MICP can be used as a 'bio-grout' to strengthen soil, instead of chemical grouting (Anbu et al., 2016). The bio-grout cements material by precipitating calcium carbonate between the particles of granular material. The substrate urea is drained through the particles with a calcium ion solution containing alkophilic bacteria. The bacteria hydrolyse the urea by using the enzyme urease and calcium as the energy source (Achal & Mukherjee, 2015). A bio-cement is formed which bonds the grains together to make a solid, and when applied to soil this increases strength and provides reinforcement. This bio-cementation process by MICP has shown to decrease the porosity and the permeability which are essential factors when controlling groundwater seepage. In addition, dryer soils and non-saturated soil environments – previously found harder to grout due to the decreased control of flow – cement homogeneously when bio-grout is used.

Bio-grouting can provide adequate strength whilst having a low impact on the environment. Interestingly, the microbial reaction happens slower than its chemical counterpart, which advantageously means that it can spread through a greater volume of soil. Per cubic meter the cost of bio-grout ranges from US$0.5 - US$9, compared to chemical grout which ranges from US$2 to US$72 (Ivanov & Chu, 2008), proving bio-grout to be the cheaper option (Kawasaki & Achal, 2016).

In addition, the naturally occurring soil bacteria has the potential to be used in MICP. Until recently, most geochemical and geotechnical processes have been assumed abiotic, in part due to the relatively long-time scales under normal subsurface conditions. However, these geotechnical reactions can be amplified considerably with the use of microbial catalysts (Kawasaki & Achal, 2016). If the process is engineered, the potential of these processes can be harnessed at a much faster rate and with a larger output. For example, by adding nutrient media to soil, the cell density of certain bacteria could be improved, increasing the rate of geochemical processes.

In addition, research has shown that MICP has been utilised for strengthening compacted or rammed earth. Rammed earth buildings have a lower embodied energy in comparison to conventional clay or concrete brick buildings and consist of compacted sand, gravel, silt and clay of varying proportions. The sand mix is compacted sequentially in layers in a temporary formwork. However, this type of structure does not have the same durability, safety and strength. Porter et al. (2018) experimented in improving the load capacity, moisture resistance and thermal relationships by employing MICP techniques. Dry sand was mixed with water and compressed by a rammer in 150 mm^3 rammed earth blocks. After demoulding the blocks, they were sprayed with a bacterial and calcium chloride solution on each face twice a day for 24 days. Figure 2-3 below shows the resultant rammed earth blocks after treatment.

Figure 2-3: Triplicate rammed earth blocks after MICP treatment.

The MICP treated blocks achieved compressive strength between 0.5 and 0.8 MPa as opposed to the cement stabilised blocks which achieved strengths between 5.2 and 14.9 MPa. The surface treatment with MICP decreased the water absorption by 24%, higher than the cement stabilised rammed earth brick which only achieved a decrease in water absorption of 20%. In addition, when sprayed with high-pressure water the MICP treated bricks showed an increased erosion resistance in comparison to conventional rammed earth blocks. This increase was demonstrated by an increased in time to failure.

2.5.4 Masonry Product Improvement

Dhami, Reddy & Mukherjee (2012) experimented with enhancing the durability of ash bricks (rice husk ash (RHA) and fly ash) using MICP. They tested 30% rice husk bricks, 30% fly ash bricks and conventional red bricks. The bricks were soaked in a tub of *Bacillus megaterium* culture for four days. When the bricks were taken out, they were cured over four weeks by spraying on NBU (nutrient broth media) and a urea $CaCl_2$ cementation solution. After four weeks, the bricks were dried in an oven at 50°C. They were then tested for water absorption, compressive strength and freeze-thaw resistance. The results showed that the bacteria effectively deposited calcium carbonate on the surface of the ash bricks, leading to a reduction in permeability (enhanced durability) and an increased compressive strength.

Furthermore, Mukherjee et al. (2013) conducted tests on MICP as a process for improving the performance of low embodied energy, soil-cemented bricks. Soil-cemented bricks have adequate strength, but they absorb high levels of moisture. In humid conditions, they become soft and non-uniform expansion causes cracking and deformation. *Bacillus megaterium* was used as a curing spray for a period of 28 days after which a significant reduction in water absorption and porosity was observed, along with an increase in compressive strength. Both studies suggest that a surface treatment that stimulates bacterial activity can create a barrier layer that can improve durability and strength, allowing a larger scale of use in industry.

2.5.5 Bio-cementation

Bio-cementation is the MICP process to be applied in this study. It focuses on cementing loose material with no prior cementation. Duo et al. (2018) investigated solidifying Aeolian desert sand using MICP. The study used varying concentrations of solidification solutions (0.5, 1.0, 1.5, 2.0 and 2.5 mol/L urea-CaCl$_2$) and the cylindrical plexiglass tube set-up as shown in Figure 2-4.

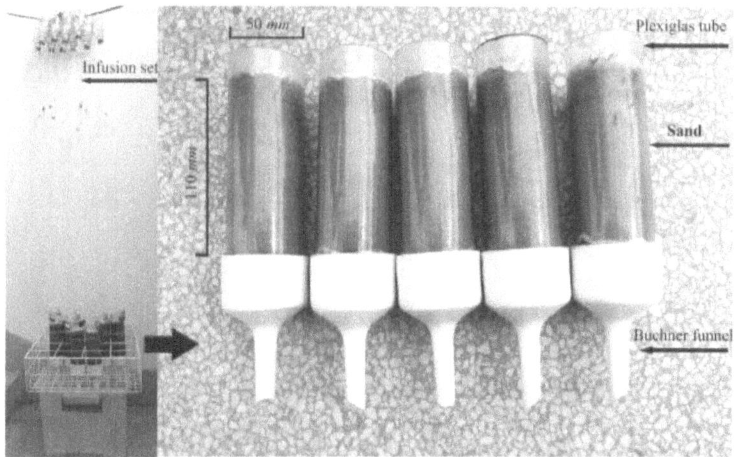

Figure 2-4: Testing mould used for bio-cementation by Duo et al. (2018). The Plexiglas tubes are connected to a funnel at the bottom. The funnel is filled with thick gravel layer with a non-woven geotextile on top of the gravel to prevent seepage of sand. Another layer of geotextile was then placed on the top of the sand column to prevent the erosion of the surface sand (Duo et al., 2018).

The results showed that adding the bacteria culture and cementation solution to the moulds solidified the Aeolian sand into a bio-solid as shown in Figure 2-5. In addition, increasing the concentration of the cementation solution, increased calcium carbonate precipitation resulting in increased sand density, decreased permeability and increased unconfined compressive strength (UCS). Duo et al. (2018) also mapped the relationship between the calcium carbonate (CaCO$_3$) content and the UCS and showed a positive correlation. The microstructure of calcium carbonate crystals were analyzed under an optical microscope and were found to be rhombic hexahedrons that wrapped around the sand particle surfaces and filled the pores. Also, the size of the calcium carbonate crystals increased from 5–8 µm to 10–14 µm as the concentration of the cementation solution increased.

Figure 2-5: The appearance of solidified aeolian sand specimens taken from solidifying test (a) and the specimen undergoing an unconfined compressive strength (UCS) test (b).

2.5.6 Bio-MASON: Bio-solids from Urea

The most notable application of MICP, with specific relation to the objective of this work, was implemented by the company BioMASON. BioMASON is a company based in North Carolina (USA) who use microorganisms to grow masonry products. The bio-bricks are produced by mixing microorganisms in a unit mould and then fed an aqueous solution to harden the bio-bricks to specification. They use synthetic urea to enable MICP to take place. BioMASON have shown that the process can be scaled to a manufacturing standard by utilising traditional casting methods; or it can be done digitally using computer numerical controlled (CNC) technologies, to programme material compositions to fabricate layered units. This allows the company to customise the material distribution, consistency and form of each unit. The programming of layered growth means they can vary the dimensions within the bio-brick. They compare this to how a bone varies in density and orientation throughout its length (Dosier, 2016). It is possible for the bio-bricks to contain a network of holes to reduce overall weight but still maintain support strength. Their bio-bricks have been found to have a greater strength than traditional clay bricks.

Figure 2-6 details the combination of the rotational and lamination methods which BioMASON employs to manufacture their bio-bricks. In step 1, *S. pasteurii* is grown in a yeast extract or peptone. In step 2 the aggregate is added into the formwork. If employing the lamination method, a thin layer of loose aggregate is added. In step 3 an aqueous solution is fed through the aggregate material (Dosier, 2016). The solution contains urea and calcium ions obtained by adding a salt like calcium chloride. In step 4 the enzyme broth solution is fed through the or done simultaneously. Calcium carbonate crystals are formed between the pores of the aggregate material in step 5. In step 6, the solution used in step 3 is used multiple times until the brick reaches its desired hardness. The formwork is rotated in step 8 and steps 3 to 7 are repeated (Dosier, 2016). Once the layer has reached a desirable strength, a new layer is added on top of the hardened one. Steps 3 to 8 are repeated for each additional layer. Cementation is more consistent, and the material structural performance is improved by using the layering method (Dosier, 2016).

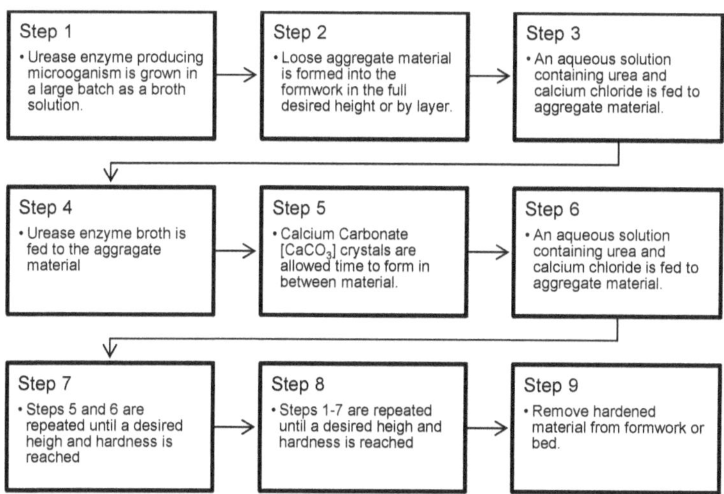

Figure 2-6: BioMASON's bio-brick manufacturing process (Dosier, 2016).

2.5.7 Bio-solids Produced from Synthetic Urine

(Henze & Randall, 2018) investigated the use of MICP for the production of bio-solids using synthetic urine. This was tested to see if urine could be potentially used as an alternative source of urea for bio-brick production, to enable a more water and energy efficient process. The process depended on using stabilised urine, thus the ability of the bacteria (*S. pasteurii*) to survive in a high alkaline environment needed to be investigated. Henze & Randall (2018) investigated how the alkalinity of the synthetic urine affected the ureolytic activity of the bacteria. The results showed that MICP could operate in a synthetic urea solution with a pH of 11.2, allowing stabilised urine to be used as a cementation media in the MICP process.

Column moulds shown in Figure 2-7 were used for the MICP process in work by Henze & Randall was pumped through the column. The cementation media was pumped through to fill the pore volume, with a retention time of three hours. The pH of the cementation media was adjusted with 0.1 M of HCl (Henze & Randall, 2018) in accordance with a titration conducted on the synthetic urea. Calcium chloride was added to increase the influent calcium concentration in order to maximise the precipitation of calcium carbonate between the pores of the masonry sand.

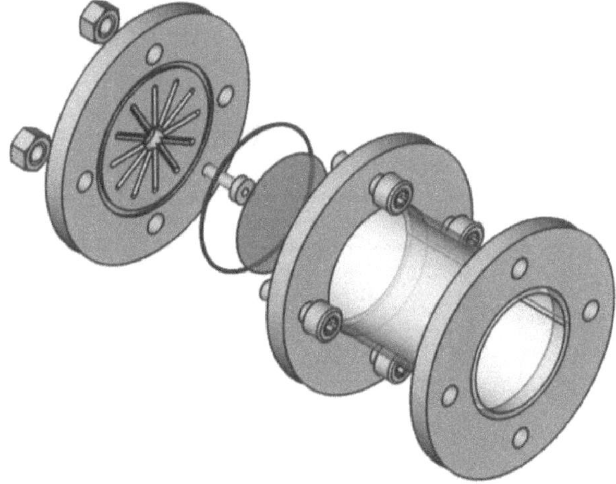

Figure 2-7: The columns moulds used to make bio-solids. The column moulds have a volume of columns 255 mL, with an inside diameter of 57 mm and a length of 100 mm. The bottom and top lids are identical and removable, both sealed with an O-Ring. On the inside face of both lids, a radial depression distributes the incoming liquid over a large part of the sand body surface (Henze & Randall, 2018).

Bio-solids formed during the study can be seen in Figure 2-8 .

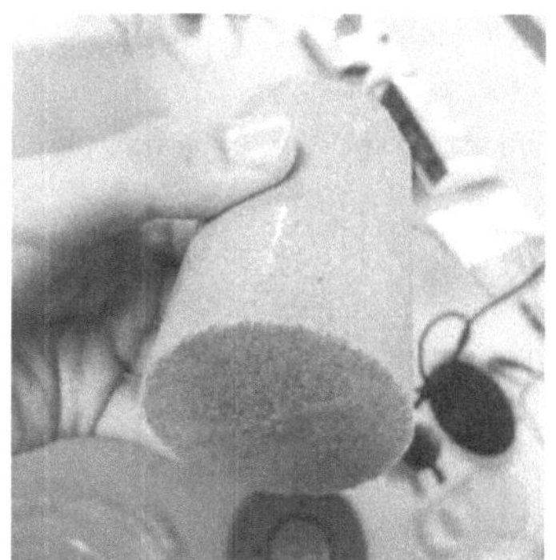

Figure 2-8: Bio-solids formed in the column moulds using synthetic urine (Henze & Randall, 2018).

2.5.8 Bio-solids from Pig Urine

Pig urine causes environmental pollution due to the high pig urea (PU) content in its waste disposal (Chen et al., 2018), but the biotechnical use of it can contribute to alleviating its own environmental impact. Chen et al. (2018) investigated inducing MICP using PU as an alternative to industrial urea. *S. pasteurii and* PU mixed 1:1 with calcium chloride ($CaCl_2$) made up a cementation media which was injected into quartz sand columns. PU was obtained from centrifuging pig urine collected from a pig farm nearby. Interestingly, the research reported nothing on issues regarding undesired urea hydrolysis between collection or treatment. It mainly concluded that using PU allowed for MICP and enhanced mechanical properties of the columns, providing a cheaper option to conventional industrially produced urea, alleviating costs and environmental pollution. However, this current study focuses on utilising human waste streams and not animal alternatives as a source of urea in the application of MICP.

2.5.9 Bio-solids from Human Urine

Lambert (2017), using the same equipment and methods as Henze & Randall (2018), conducted the tests using real urine. The study showed that bio-solids could be produced using real urine and, theoretically, assuming the calcium usage efficiency is 85%, it would take 8 days and 56 treatments to fill a pore volume of 12.5% allowing the column to solidify (Lambert, 2017). It was hypothesised that to produce bio-solids with the volume of a standard brick, the experimental inputs required: 1.72 dm^3 masonry sand, 36 L of urine, 433 g of calcium chloride dihydrate, 133 mL of 32% HCl and 850 mL of bacteria culture.

Further testing of the mechanical and physical properties of bio-solids formed using MICP from are comparable to construction material used in industry (Mukhari, 2018). Investigations on the compressive strength, porosity, water absorption and calcium carbonate microstructure at different stages of the treatment cycle were carried out. The results indicated that a bio-solid of adequate strength could be grown with a maximum strength of 5.1 MPa (Mukhari, 2018). The Scanning Electron Microscope (SEM) images showed the nature and structure of the calcium carbonate crystals formed and its polymorphs. It also showed that the calcium carbonate crystals increased in size with increasing treatment time (Mukhari, 2018). The method used to grow the bio-solids focused on how many treatments of cementation media were required for solids to form, whilst keeping the influent calcium constant at 0.11 M. The experiments showed that the minimum time taken to form a solid when the bacteria-inoculated sand mixture, was 27 treatments, amounting to 4 treatment days in total (Mukhari, 2018).

2.6   Calcium Carbonate Precipitation

Calcium carbonate can exist in three polymorphs: calcium carbonate, aragonite and vaterite (Mukhari, 2018), with calcium carbonate being the most thermodynamically stable and vaterite being the least (Dickinson & Mcgrath, 2004). Wang et al. (2013) showed that temperature has a role to play in which form calcium carbonate precipitates; all three (Mukhari, 2018) crystal polymorphs including amorphous particles were found at 25°C whereas at 55°C calcium carbonate and aragonite were most prominent (DeJong, Fritzges & Nüsslein, 2016).

Figure 2-9 shows the three arrangements that calcium carbonate can form around loose sand (Lin et al., 2016): (a) shows the precipitate forming around grain contact points; (b) shows the precipitate forming inbetween the pore spaces without coming into contact with the grains themselves and (c) shows the precipitate forming in the pore space and on the grains, bridging the grains together (matrix supporting) (Lin et al., 2016).

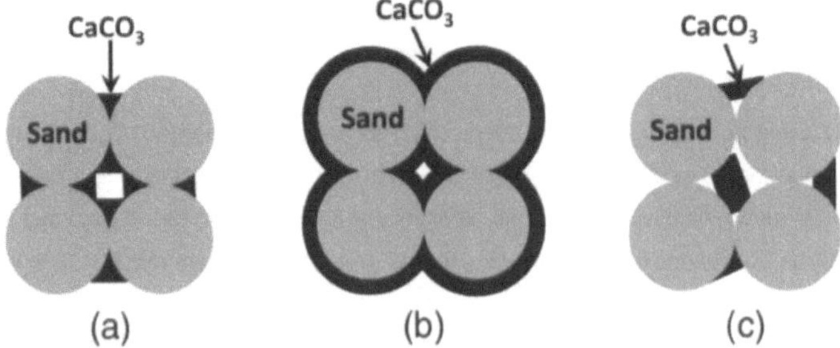

Figure 2-9: Arrangement of calcium carbonate in sand pore space during MCIP: (a) contact cementing, (b) grain coating, (c) matrix supporting (Lin et al., 2016).

The arrangement the precipitate in the MICP process can indicate the strength of a solid produced. The pore volume between grains is critical to such a process. If the pore volumes are small, less calcium carbonate will precipitate, and if the pores are too small, it may inhibit the movement of bacteria which promote cementation (DeJong, Fritzges & Nüsslein, 2016). Conversely, if pore volumes are too large then the compressive strength is compromised (Chen, Wu & Zhou, 2013).

Calcium carbonate can only precipitate on and around sand grains that are in contact with bacteria (Mukhari, 2018). Depending on the rate of urea hydrolysis, calcium carbonate can form as a crystal or an amorphous shape. Whiffin, Van Paassen & Harkes (2007) showed that high hydrolysis rates produced spherical and amorphous precipitates, whereas low urea hydrolysis rates produced rhomboidal shaped precipitates. In essence, changing the rate of hydrolysis has the potential to directly affect the strength properties by manipulating the precipitate formed at a microscopic level (Mukhari, 2018).

The supersaturation level of carbonate ions can be further influenced by controlling the pH. At a higher pH the carbonate ion concentration is greater, meaning that if you lower the high pH instantly, the size of the crystals that form will be bigger (DeJong, Fritzges & Nüsslein, 2016).

2.7 Compressive Strength

Compressive strength informs what load a material can withstand and is used worldwide as a comprehensive indicator of a material's strength properties. The compressive strength is the ratio of the maximum uniaxial load applied to the cross-sectional area of the object (Mukhari, 2018). For concrete, the South African National Standards (SANS) tests this property using cube compression tests, while the Eurocode uses cylindrical specimens. For the bio-solids, it is critical to test this marker according to conventional practices to compare with current industry materials.

Test procedures indicate that brick specimens are to be tested flatwise and centred under the spherical upper bearing (Alam, 2018). The load is to be applied up to one half of the expected maximum load, at any convenient rate. After that, the remaining load has to be applied at a uniform rate over 1-2 minutes (Alam, 2018).

Figure 2-10: Brick testing procedure (Alam, 2018)

The calculation for compressive strength is:

$$C = \frac{F}{A}$$

2-8

Where C is the Compressive strength (MPa), F is the calibrated maximum load in Newtons, and A is the average gross area of the upper and lower bearing surfaces of the specimen measured in square millimetres (Alam, 2018).

2.8 Sporoscarcina Pasteurii

The bacteria, *Sporosarcina pasteurii (S. pasteurii),* is a common soil bacteria which is non-pathogenic (Anbu et al., 2016) and a picture of the bacteria under the microscope can be seen in Figure 2-11. This bacteria can fuse loose aggregate by inducing the formation of cement material. This can be done between loose grains of sand, soil, fibres, basalt, glass beads, recycled glass foam, fly ash composite, soil and small stones. In addition, any of these can be mixed and subsequently fused. This strain of ureolytic soil bacteria is an endospore producing bacteria. It has an optimal growth at a pH of 9 with an ability to tolerate extreme conditions (Anbu et al., 2016, Henze & Randall, 2018). It is also robust in a high alkaline environment (Henze & Randall, 2018). The organisms are chemoheterotrophic, aerobic, gram-positive rods. They are 2-3 µm in length, are basophilic, form round spores (diameter 0.5-1.5 µm) and show an optimal growth at 25-37°C (Duo et al., 2018). Heterotrophic means that it oxidises the carbon found in organic compounds (plant or animal matter) as its principal energy source and cannot produce its own out metabolic processes and subsequently can be subdivided into a chemoheterotroph. The bacterium is aerobic and undergoes respiration, in which adenosine triphosphate (ATP) production is coupled with oxidative phosphorylation, leading to the release of oxidised carbon wastes such as CO_2 into the atmosphere (Bingle, 2016). This CO_2 is converted into carbonate ions during MICP processes.

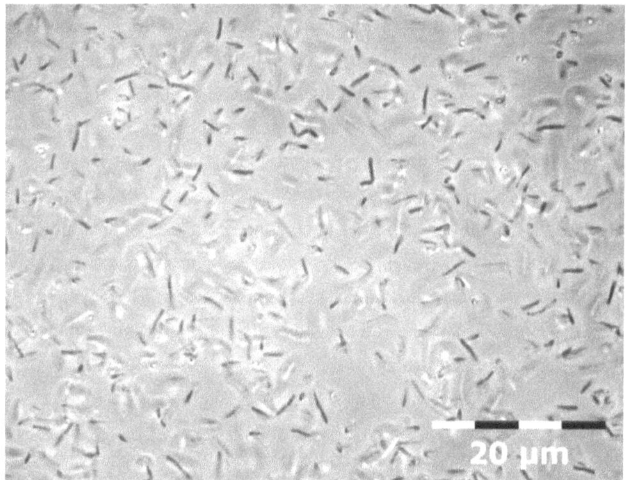

Figure 2-11: An image of *S. pasteurii* under the microscope (Lambert, 2017).

The pH of the nutrient medium has been shown to affect the biomass growth of the bacteria. Work by Cuzman et al. (2015) showed that increasing alkalinity caused the lag phase and exponential phase of bacteria growth to be extended and reduced respectively because of the stress on the bacteria. A decrease in the production of ATP was observed at an alkalinity of 12, but the surviving bacteria seemed to adapt after 4 hours and subsequently grew again. The biomass produced was half that of the maximum urease activity which occurred at a pH of 7. The experiments conducted by Cuzman et al. (2015) showed that in extreme conditions the urease hydrolysed per minute decreased and that the alkalinity had a greater effect if the starting concentration of the bacteria was low.

2.9   Urea hydrolysis, Urine Stabilisation and Value Recovery from Urine

To achieve the objective of this study, to use human urine as a source of urea for MICP, the urine must be stabilised to prevent the loss of urea by enzymatic urea hydrolysis. There are a number of ways to stabilise urine, but for the purpose of this book, we will only look at and so it is ideal for incorporating a technology that both recovers inorganic fertilisers and stabilises and prepares the urine for the requirements of this study. Conversely, the urea can be extracted from fresh urine by concentrating it to be used later as a diluted solution for MICP.

2.9.1 Urea Hydrolysis

In urine, approximately 90% of nitrogen is in the form of urea (Henze & Randall, 2018). The enzyme urease is a catalyst for urea hydrolysis and the reaction results in the release of CO_2 and free volatile ammonia. The opportunity to recover nitrogen fertiliser from urine is lost with the release of free volatile ammonia into the atmosphere. Enzymatic urea hydrolysis and chemical urea hydrolysis are both processes by which urea is degraded. The enzyme urease that facilitates enzymatic urea hydrolysis is prevalent in the environment and hydrolysis is inevitable in an open-air system (Randall et al., 2016). Chemical urea hydrolysis is strongly dependant on the pH and temperature of the solution. At a pH of above 13 and below 2, an increase in temperature will result in chemical hydrolysis (Randall et al., 2016). Varying pH's, high temperatures between 66°C and 100°C also affect the stability of urea and can cause chemical hydrolysis within days or hours (Warner, 1942). Figure 2-12 displays the pH and temperature in the alkaline region in which urea is not susceptible to urea hydrolysis.

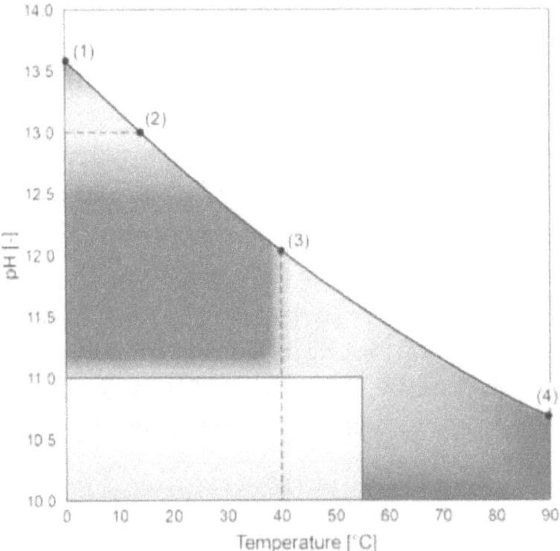

Figure 2-12: Design chart showing areas where negligible urea loss occurs (green), enzymatic urea hydrolysis occurs (bottom rectangular region, between pH 10 to 11 and temperature 0–55°C and where chemical urea decomposition is more abundant (yellow–orange-red). Red being the region where the greatest urea loss will likely occur (Randall et al., 2016). The dotted lines represent precautionary limits for chemical urea hydrolysis, while the white area above the curve approximates the limits for enzymatic urea hydrolysis (Randall et al., 2016). The saturation pH curve for calcium hydroxide is represented by line 1–2 : 3–4 (Randall et al., 2016).

In order to avoid chemical urea hydrolysis at high temperatures, the operating temperature should be kept below 40°C (Randall et al., 2016). Enzymatic urea hydrolysis can be prevented by electrochemical treatment, the addition of natural and artificial inhibitors, a pH decrease by the addition of acid or a pH increase by the addition of calcium hydroxide or wood ash (Randall & Naidoo, 2018). If both the enzymatic and chemical hydrolysis are inhibited, then urea is available for recovery as a stable crystal form. The section below explores research that informs the viability of applying these technologies in the bio-brick production proposed in this book.

2.9.2 Enzymatic Urea Hydrolysis Inhibitor Technologies

The following technologies were analysed to inform which method of urea hydrolysis inhibition were most suitable for this study. The desired method had to adequately inhibit urea hydrolysis at room temperature, produce a liquid that could be pumped through the bio-brick process, not precipitate undesired solids during MICP, be safe to handle and also be affordable. Finally, it should allow for the recovery of fertiliser before and after MICP.

A study was done by Ikematsu et al. (2007) to test whether electrochemical treatment of human urine could be used to suppress the activity of urease to enable its further reuse as flushing water. The inhibition of urea hydrolysis is caused by an irreversible change in urease's conformation, achieved through oxidation by electrochemically produced chlorine. Its irreversible inactivation occurred at or above an oxidation-reduction potential (ORP) of 240 mV. The urine was electrochemically treated by sandwiching a platinum electrode between two platinum-iridium electrodes. However, this kind of treatment only allowed for short-term inhibition, and future urea hydrolysis from contact with urease-active microorganisms was not prevented (Ikematsu et al., 2007). This method also did not facilitate the recovery of fertiliser(s).

Another study focused on using allium (garlic, onion, leek) and brassica (cabbage, brussel sprouts) plant juices to inhibit enzymatic urea hydrolysis (Olech, Zaborska & Kot, 2014). The enzyme urease is widely found in nature; however, bacterial and plant ureases differ in their molecular structure. This phenomenon allows for plant ureases to serve as inhibitors, while bacterial ureases function as catalysts for urea hydrolysis. The ability of the plant juices to inhibit urea hydrolysis was directly correlated to their Thiosulfinate (TS) content. Garlic juice was found to be the most efficient inhibitor and reduced the initial urease activity by 60-75%. To make the plant juice, the various plants had to be sliced, pulped, filtered through a gauze and centrifuged. The inhibition of urease activity is still too unreliable, with less than desired inhibition ability for the proposed bio-brick process.

A technique to alkalise and inhibit enzymatic urea hydrolysis by means of adding wood ash at 35°C and 65°C was explored by Senecal & Vinnerås (2017). The technique used a container-alkaline ash-based dehydrating bed. This system preserved 90% of the nitrogen and saw a 95% volume reduction during dehydration. The volume reduction can allow for an ease and reduction in transportation and storage costs while enabling the final product to be used as a fertiliser. The final product was in the form of a dry powder, and by weight, it contained 7.8% N, 2.5% P and 10.9% K, which are comparable to values of commercial fertilisers and even more concentrated than manures (Penhallogen, 2003; Senecal & Vinnerås, 2017). The wood ash allowed the urine's pH to be alkalised to 10.3-10.7, and any losses of nitrogen were found to be as a result of temperature rather than pH. This method would require dissolving the product as needed to put the urea in a liquid form to implement in the pump system used in the bio-brick production which would require unnecessary amounts of water to dilute the crystallized urea when eventually used for MICP.

Hellström, Johansson & Grennberg (1999) found a strong relationship between pH and the amount of urea decomposed in the urine. Their research showed that at the beginning of a storage period, the decomposition of urea could be avoided by dosing a litre of undiluted urine with 60 meq of sulphuric or acetic acid. Although only two acids were tested, the research recommended using sulphuric or phosphoric acid, which could also produce fertilisers. However, phosphorous ions aren't precipitated out and can, therefore, have the undesired effect of precipitating inside the bio-brick system. In addition, handling acid is unsafe, and a pumping mechanism would be required to implement such a system.

Randall et al. (2016) found that strong bases in the form of calcium oxides and hydroxides can also stabilise urine. The calcium forms phosphates at high pH's during which pathogens are also simultaneously destroyed. Calcium hydroxide is a cheap base often used for sanitation purposes, and it precipitates calcium phosphate during stabilisation (Flanagan & Randall, 2018). The calcium phosphate can be recovered for use as a fertiliser (Meyer et al., 2018). As a result, the stabilisation of urine with calcium hydroxide becomes a multi-faceted process which both recovers nutrients and prevents enzymatic urea hydrolysis. As opposed to other acids, it is safe to handle and can be used in a passive dosing system to boost the pH of soils, which is why it is used commonly in agriculture (Flanagan & Randall, 2018). Calcium phosphate is also recognised as a particularly effective crop fertiliser; it is highly soluble and does not disrupt the pH levels in the soil. It can also bond with calcium, which is naturally present in the soil, to form dicalcium phosphate that serves as additional crop nutrition (Flanagan & Randall, 2018). Maintenance on waterless urinals is often difficult because of scaling caused by phosphate crystals. By harvesting phosphates as a fertiliser, the need for maintenance is reduced. It was found by Flanagan & Randall (2018) that the precipitation of calcium phosphate from fertiliser-producing urinals was possible. It can be concluded that dosing urine with calcium hydroxide is a preferential method for stabalizing urine for bio-brick production since:

- No active dosing is required;
- It produces fertilisers and ensures the precipitation of phosphates from the liquid before being fed through the bio-brick production system;
- The process allows for the urea to remain in a liquid solution which can then be pumped through the bio-brick system;
- Calcium hydroxide is an easily available, safe to handle and cheap base;
- The resultant high pH likely ensures that bacteria and viruses are killed (Randall et al., 2016).

2.9.3 Concentrating Urea

Urea has a high saturation point of 1.21 kg/L at room temperature (Lee & Lahti, 1972) but the urea concentration within urine is typically very low, never exceeding 24 g/L (Randall & Naidoo, 2018). Therefore, to extract a crystallised form of urea, 98% of the water would need to be evaporated and would have to occur below 40°C to avoid chemical urea hydrolysis. This would result in low evaporation rates (Randall & Naidoo, 2018). In addition, as discussed in 2.5.8, concentrated urea can be obtained from centrifuging pig urine (Chen et al., 2018), but this research did not discuss in detail any effect of enzymatic urea hydrolysis, before or after centrifuging. Little research has been published on concentrating urea, and a more in-depth study on the specific conditions to which it could work would have to be investigated and established for these methods to be feasible for the isolation of urea. Furthermore, both concentration methods would require unnecessary amounts of water to dilute the crystallized urea when eventually used for MICP.

2.9.4 Dosing with Calcium Hydroxide

Testing the stabilisation method of dosing urine with calcium hydroxide has been conducted in a study by Randall et al. (2016) which recommended a dosage of 10 g/L of calcium hydroxide to ensure saturation for a wide range of urine compositions. Randall et al. (2016) also showed that when fresh urine was dosed with 10 g/L of calcium hydroxide at 25°C, the phosphate and magnesium ions were almost completely removed from the liquid phase increasing the dissolved calcium concentration within the urine. The feasibility of the urine stabilisation method of calcium hydroxide dosing was investigated in Flanagan & Randall (2018), where a portable nutrient recovery urinal showed that the urinal not only stabilised the urine but also produced a calcium phosphate fertiliser on-site. A 25 L waterless urinal system was operated for three weeks. The urinals were pre-dosed with calcium hydroxide such that when full, the concentration would be 10 g/L. On average the urinals were filled every 3.75 days, and once filled they were left to stand so that the precipitated calcium phosphate could settle. Thereafter, the liquid and solids were separated by pumping out the liquid supernatant. The solids were dried in trays at ambient room temperature (Flanagan & Randall, 2018) and the amount of solid remaining could then be measured.

During the three weeks of operation, approximately 1.1 kg of solid fertiliser was collected from 100 L of urine. A rough calculation was made to understand the significance of this study better. The University of Cape Town (UCT) has 260 standard urinals on its upper campus. If each of these urinals were retrofitted with a container for urine collection and fertiliser production, the amount of fertiliser that could be collected per day was estimated to be 73 kg (Flanagan & Randall, 2018).

Additionally, using waterless urinals offered indirect benefits such as water savings. The waterless nutrient recovery units proposed by Flanagan & Randall (2018) does not require any water for flushing, and this is beneficial as there is no dilution of the nutrient-rich urine, resulting in a higher yield per litre of nutrients recovered. Another advantage of a waterless system is its relevance in a water-stressed country like South Africa, where strict daily water restrictions have to be adhered to in some areas. A study by Chipako (2019), shows that an average urinal at the University of Cape Town (UCT) is visited about 75 times per day. This equates to 165 L/day of water saved in just one of these urinals. Phosphate could be collected as calcium phosphate during stabilisation which also means a decreased demand for cleaning and maintenance as calcium phosphates often precipitate as solids in urinals and pipes system causing scaling and blockages.

On average, conventional urinals use up to 4 litres of water per flush (von Munch et al., 2009). Implementing water saving devices could reduce this volume substantially. Chipako (2019) conducted a hypothetical case study on the feasibility of implementing waterless nutrient recovery urinals at the University of Cape Town. Chipako (2019) found that 19 ML of water could be saved every year from not flushing conventional urinals. This translates into a direct operating cost saving for buildings. Avoiding flushing water also reduces the cost of treatment as the volume of urine requiring treatment is less. Additionally, potable water is often used for flushing toilets and is subject to ever-increasing shortages. Therefore, waterless urinals should be installed in all buildings where possible. Moreover, in Chipako's study, he conducted an economic evaluation on implementing resource recovery urinals at UCT (Chipako, 2017). One scenario focused on outsourcing the installation of waterless urinals, their maintenance and treatment, to an existing company. Another scenario based on the business model on the university providing the service internally. Chipako (2019) showed that a revenue of ZAR 120 000 could be obtained from the annual production of fertilisers 260 urinals at UCT, amounting to 6700 kg. If outsourced, that revenue would be obtained by the external company. If insourced, the capital costs would be ZAR 6 million and the operating costs per annum would be external company shows itself to be a better option than the current conventional system, as it does not require sizeable capital input but still promotes water saving and resource recovery. Additionally, outsourcing would create a market demand for the integrated nutrient recovery system. This analysis could also be applied to office buildings.

Chapter 3 Materials & Methods

This chapter describes the methods of microbiological culture and growth as well as urine stabilisation used in this book (section 3.1 and 3.2 respectively). Sections 3.3 – 3.9 further describes the methods used to achieve the research objectives. Section 3.3 describes the methodology pertaining to the feasibility and optimisation of the bio-brick production system. Section 3.4 – 3.8 describe side experiments which singled out and tested certain factors to observe their effect on the bio-brick system. The factors include ionic strength, pH and calcium concentration of the influent cementation media. Lastly, the chapter ends by describing the experiments conducted on alternative nutrient media for the bacteria strain used in this book.

3.1   Micro-biological Culture and Growth

S. pasteurii was grown according to the protocol described by (Bhaduri et al., 2016) . Ammonia-Yeast media (ATCC˚ 1376) was used for the propagation of *S. pasteurii* as either a bacteria culture or poured onto agar plates which consisted of ammonium sulphate, yeast extract and tris buffer. To produce ATTC˚1376, deionised water was mixed with tris-base to make a 0.13 M aqueous solution of tris buffer. The pH of the tris buffer was then lowered to a pH of 9 using HCl the solution was then divided into two parts to avoid the milliards reaction from occurring; 10 g/L of ammonium sulphate was dissolved in one of the tris-base solutions, and 20 g/L of yeast extract was dissolved in the other. The solutions were then autoclaved separately as the milliards reaction occurs at high temperatures between amino acids and sugars. After the two solutions were autoclaved, they were mixed and then kept in a sealed bottle and stored in a fridge.

The agar plates were made according to the method employed above to make the Ammonia-Yeast media (ATCC® 1376), but before autoclaving, 1:1 agar was added to the yeast solution. After autoclaving, the two solutions (the ammonium sulphate solution and the yeast and agar solution) were mixed and then poured into petri dishes. This was done immediately after autoclaving to ensure the agar did not solidify. The petri dishes were then sealed using laboratory film and stored in a fridge.

As seen in Figure 2-1 below, to grow the bacteria, a bacterial stock was removed from the freezer (-50°C) and allowed time to thaw. In an aseptic environment under a fume cupboard, a heated loop wire was dipped into the frozen stock. The bacteria on the loop wire was then plated onto the agar plates. The plates were then stored in an incubation room at 30°C to grow, and after 48 hours the plates were examined for the presence of colonies.

Once colonies were present on the plates, 100 mL of ATTC®1356 was poured into 500 ml Erlenmeyer flasks in an aseptic environment. A heated loop wire was used to gently remove colonies of bacteria from the agar plate into the ATTC®1356 solution. The Erlenmeyer flasks were then placed in an incubator at 30°C at 150 rpm for 2-3 days.

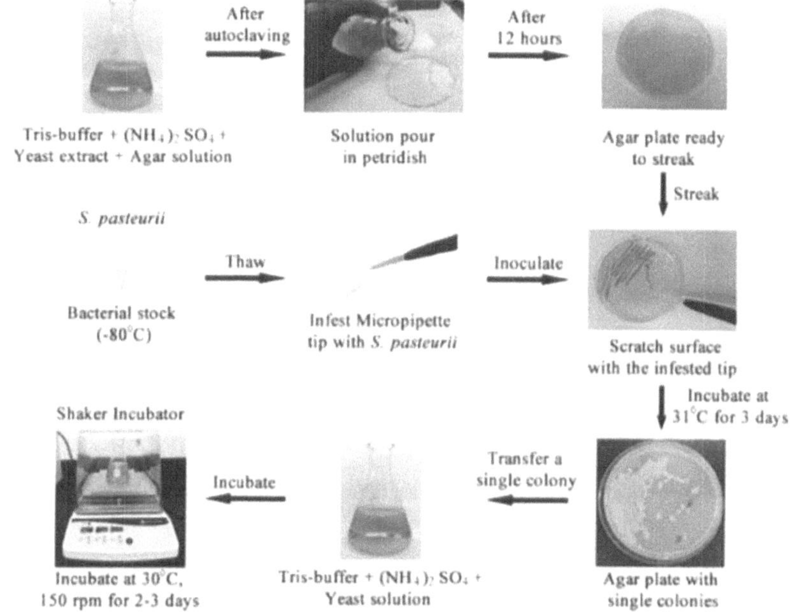

Figure 3-1: A Figure to show the protocol followed to grow *S. pasteurii*. (Bhaduri et al., 2016).

Afterwards, the bacteria culture was then sub-cultured into three separate Erlenmeyer flasks, with 100 mL of ATTC®1356 at a starting optical density (OD) of 0.05. The inoculum was then left overnight in the incubator and harvested at about 15 hours; the late exponential phase according to the growth curve produced by Lambert (2017) in Figure 3-2.

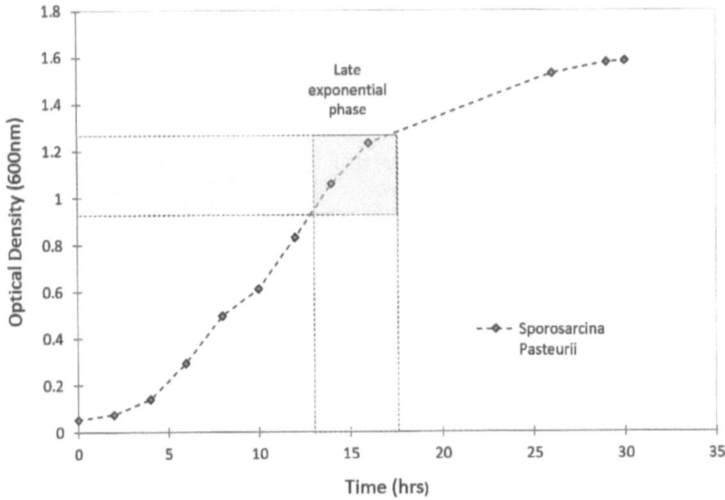

Figure 3-2: A Graph showing the period where the bacteria was harvested for use in MICP experiments (Lambert, 2017).

3.1.1 Glycerol Stocks

Glycerol Stocks were prepared for storage to be used later. An Erlenmeyer flask filled with 100 ml of ATTC®1356 media, was inoculated with 5 mL of a bacteria culture at stationary phase, resulting in an OD of about 0.1. The flask was then incubated until mid-exponential phase at which point a 1 mL sample was extracted. The sample was aseptically mixed with 1 mL of autoclaved 50 vol% glycerol and poured into sterile, cryogenic vials to a volume of 2 mL (Zhang, 2013). The vials were then stored in a -50°C freezer.

3.1.2 Concentrated Culture

A 100 mL of bacteria culture was grown overnight in ATTC®1356 media. Once it had reached the late-exponential phase the bacteria culture was split into two 50 mL tubes. The tubes were then centrifuged at 2570 g for 20 minutes. The centrifuging caused a concentrated bacteria pellet to form at the bottom of the tubes. The supernatant was then decanted carefully using a pipette until only 10 mL of the surrounding liquid remained. Using the pipette, the pellet was re-suspended and mixed with the remaining liquid in order to concentrate and multiply the cell density. The two 50 mL tubes were mixed, and the OD was measured.

3.2 Urine collection

3.2.1 Urinal Design

Urine was collected according to the methods and materials described in Flanagan & Randall (2018). The components of the nutrient recovery urinal (NRU) can be seen in Figure 3-3. A funnel was attached to the top of the urinal which was used to guide the urine into the container below, and a backsplash cover was added to avoid the urine spraying out of the funnel. The lower half of the system was a 25 L container that collected the urine and provided a space where it could mix with the pre-dosed calcium hydroxide. A threaded PVC pipe connected the collection tank with the funnel so that the end of the funnel extended further than the containers threading, thus ensuring a clean system. The funnel was removable so that the container could be closed and sealed using a normal lid. This aided in making the system sanitary and the urine storable.

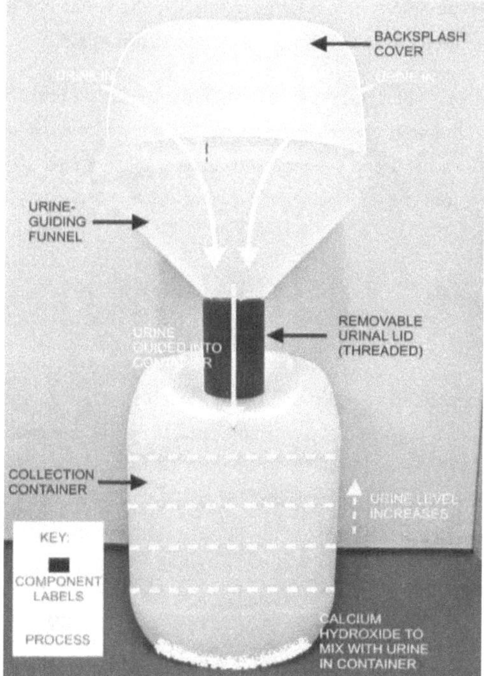

Figure 3-3: A Figure to show how the nutrient recovery urinals work (Flanagan & Randall, 2018).

3.2.2 Urine Collection

The urinal was installed in a male bathroom situated on the third floor of the New Engineering

Environment (EBE) ethics requirements with a poster informing urine donators on the experimental use of their urine and their right not to participate if they so wish.

The urinal was pre-dosed with 250 g of calcium hydroxide. This was done to ensure that there was 10 g of calcium hydroxide per litre of urine (Flanagan & Randall, 2018). The urinals were monitored every day to avoid the collection containers overfilling. Monitoring also encompassed 'swirling' the collection containers every evening to ensure that the urine was mixed with the calcium hydroxide such that a high solution pH was maintained.

3.2.3   Urine Processing

Once the collection tank was filled, it was removed from the bathroom, sealed and left to stand. This was necessary to allow for the calcium phosphate precipitate to fully form and settle at the bottom of the tank. It was essential to monitor the temperature of the stabilised urine to avoid saturation at increasing temperatures. Before filtration, the stabilised urine had to reach room temperature and have a pH around 12- 12.5, if not the urine was discarded.

Per experiment, approximately 33 L of urine was used to produce one bio-brick. The urine was filtered and stored before the start of each experiment. Using a peristaltic pump, the liquid supernatant was carefully removed and channelled through a vacuum filter to separate any suspended solids. A 1.2 µm pore size filter with a diameter of 150 mm was used (595 Schleicher & Schuell Dassel, Germany).

3.3   Analytical Methods

3.3.1   pH Measurement

A pH probe (HI1131B, Hanna Instruments, Rhode Island, United States) was used to measure the pH of the solutions. Before experiments were conducted, the probe was calibrated using 3 points: 7.01, 10.01 and 12.592. Hanna pH-buffers were used for the first two calibration points, and a saturated calcium hydroxide solution was used for the third. A calibration point of 12.592 was chosen for this study, which was the pH of the saturated calcium hydroxide at a measured temperature of 21°C.

3.3.2   Optical Density

The Optical Density (OD) was measured using a GENESYS 10 Photospectrometer (ThermoFisher Scientific, Massachusetts, United States). This instrument records the suspended cell biomass by measuring the returned absorbance from a light with a wavelength of 600 nm. The OD was assumed to be directly proportional to the concentration of the suspended cell biomass. A sample of ATTC®1356 media was used as the blank (zero absorbance reference).

3.3.3   Sampling Procedure

The Thermoscientific Gallery (TSG) (ThermoFisher Scientific, Massachusetts, United States) was used to sample ammonium, urea and calcium concentrations from the influent and effluent streams. The samples were diluted to comply with the reading range of the TSG. The calcium concentration range was between 10-200 mg/L, and the ammonium concentration range was between 1-5 mg/L.

The sampling procedure for calcium ions was carried out as follows: the influent and effluent samples were filtered with a 0.22 μm pore syringe filter. A sample of 1 mL of the influent and 2 mL of the effluent filtered samples were each pipetted into three graduated 15 mL vials. The vials were then topped up to the 10 mL mark with deionised water to make a dilution of 5 and 10 respectively.

The TSG is unable to measure the concentration of urea but can measure the concentration of ammonium ions. The influent urea was therefore measured by hydrolysing all the urea into ammonium, using Jack Bean urease (U1875- 25 mL, Sigma Aldrich, St Louis, United Sates), and then subtracting the initial ammonium concentration found naturally in the cementation media (Henze & Randall, 2018). The sampling procedure for the final ammonium concentration after urea hydrolysis was carried out according to the following steps: the pH of the sample was acidified with 0.1 M of HCl to acquire a pH between 7 and 8. This was done to adjust the sample into the working pH of the enzyme, Jack Bean urease, while ensuring that the nitrogen was converted into ammonium rather than into gaseous ammonia. The influent sample was filtered with a 0.22 μm pore syringe filter, and 1 mL of the sample was pipetted into a 50 mL volumetric flask. The volumetric flask was then topped up to the 50 mL mark with deionised water. From the volumetric flask, 0.5 mL was pipetted into three 15 mL graduated vials. A volume of 40 μL of urease (U1875- 25 mL, Sigma Aldrich, St Louis, United States) was pipetted into each vial. The vials were then topped up to the 10 ml mark with deionised water and left for an hour for the urea to hydrolyse.

The effluent ammonium and the naturally occurring influent ammonium were measured using the following steps: the pH of the sample was acidified with 0.1 M HCl to acquire a pH of below 3. The samples were then filtered with a 0.2 μm pore syringe filter, and 1 ml of each sample was pipetted into 50 mL volumetric flasks, which were then topped up to the 50 mL mark with deionised water. From the volumetric flask, 0.5 mL was pipetted into three 15 mL graduated vials that were then topped up to the 10mL mark with deionised water. All samples were mixed on a vortex mixer and measured in the TSG.

3.3.4 X-ray Diffraction (XRD) Analysis

X-ray diffraction (XRD) analysis was conducted to test the crystal form of the solid precipitates. This was measured by using a Phillips PW 3710/40 (PANalytical, Almelo, The Netherlands) fitted with a curved graphite monochromator. The X-rays with a wavelength of 1.542 A were produced with a PW 1830/00 generator using a copper K-α X-Ray tube, accelerating voltage of 40 kV and current of 25 mA.

3.4 Bio-brick Mould Experiments

The purpose of conducting these experiments was first to test the feasibility of producing a bio-brick from urine using the MICP process. Secondly, the experiments sought to optimise the system design regarding the optimal influent calcium concentration, retention time and number of treatments in relation to the resultant compressive strength.

The methods and materials described from 3.4.1 to 3.4.7 explain the final methodology implemented after incremental adjustments were made during the experiments.

3.4.1 Bio-brick Mould Design & Preparation

Three bio-brick moulds were used, and each one was designed as shown in Figure 3-4. The bio-brick moulds allowed for loose sand to be held in shape while the cementation media was pumped through the mould. This facilitated MICP and allowed calcium carbonate to precipitate, solidifying the sand into the shape of the mould. The system was designed as a closed system to avoid ammonia volatilisation.

The moulds dimensions were chosen according to SABS (2007) sizing. The Imperial Brick dimensions, 222 mm long X 106 mm wide X 73 mm high, are commonly used in South Africa and were the chosen dimensions for the inside walls of the bio-brick moulds. The moulds were built with transparent Perspex and the technical drawings can be found in Appendix A.

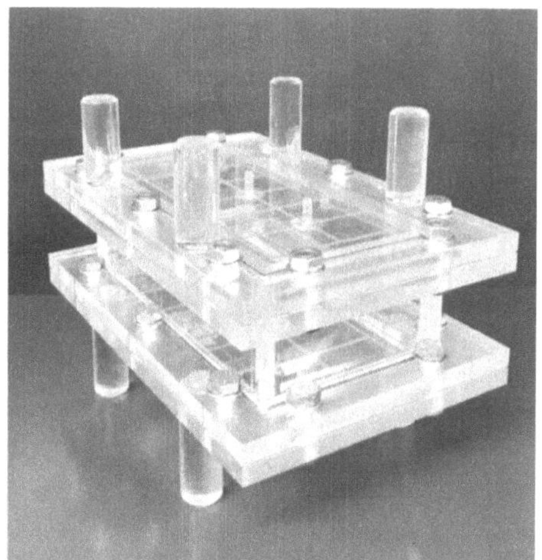

Figure 3-4: Picture of the Perspex bio-brick moulds. The bottom and top lids are identical and removable, both sealed with an O-Ring. To distribute the incoming liquid evenly over the sand body surface, the outside faces of both lids have three influent nozzles, and the inside faces have square depressions.

In preparation for the experiment, the inner walls of the moulds were lined with sheets of transparency cut to shape. To create a removable film, the space between the inner wall and the transparency was greased with Vaseline petroleum jelly. This ensured that at the end of the experiment the bio-bricks did not fuse with the inner wall and could be easily removed.

Two rectangular pieces of foam were cut, to place at the bottom and top of the mould. This put the bio-brick under a slight compressive force to avoid any preferential flow and to further disperse the incoming liquid evenly. A geotextile was placed in between the foam and the lid so that the influent and effluent nozzles did not get clogged with calcium carbonate precipitate or substrate grains.

3.4.2 Cementation Media

The cementation media is pumped through the mould at regular intervals and was prepared from filtered, stabilised urine processed in 3.2. The composition of the filtered, stabilised urine was sent for analysis for one specific batch of stabilised urine and the results are shown in Table 3-1 below. However, this is not representative of the composition of the stabilised urine for every batch used in the bio-brick experiment which would reflect a variation of values.

Table 3-1: A Table showing the typical varying concentrations of species present in stabilised urine

Species	Concentration (mg/L)
Ammonium	588
Urea	9 760
Calcium	9110
Sulphate	2000
Chloride	10500
Sodium	6040
Ortho phosphate	<0.1
Potassium	3760
Magnesium	<1.1

Nutrient broth was added to support bacterial metabolism during the experiment. Research pertaining to MICP using *S. pasteurii* suggested dosing the influent with 3 g/L of nutrient broth (Rittmann et al., 2011). After the nutrient broth was added, the media was filtered again using a Munktell Filtrak™ Micro-Glass fibre filter.

The measured mass of the stabilised urine was used to calculate the volume of cementation media. Its average density was calculated to be 0.9666 g/L. Thereafter the volume of the stabilised urine could be derived from its weight.

Based on experiments done by Lambert (2017), the target pH value was between 10.8 and 11.2. The cementation media's pH was adjusted down to the desired value by adding 0.1 M of HCl. The filtered media was then sampled for calcium, ammonium and urea concentrations.

In order to maximise the calcium carbonate precipitated, the calcium concentration of the cementation media was elevated by adding calcium chloride ($CaCl_2 \cdot 2H_2O$). The amount added was calculated from the sampled calcium concentration and the volume of the media.

3.4.3 Substrate and Pumping

Dune sand sourced from the Cape Flats and Greywacke aggregate were obtained from the CoMSIRU lab in the New Engineering Building at UCT. These were mixed to a ratio of 1:1. The Greywacke aggregate added compressive strength and increased the range of grading, which decreased the porosity, shortening the time taken to fill the pores with precipitated calcium carbonate. To avoid smaller grains exiting the system with the effluent liquid, the sand and aggregate mixture was sieved using a 0.15 mm sieve. The grains retained in the sieve where used while smaller grains were discarded.

The void ratio and the porosity of the material mixture were calculated to be 0.58 and 36% respectively. These parameters were calculated by first measuring the specific gravity of the material mixture. The specific gravity (G_s) of the soil was established using the Density Bottle Method outlined in the British Standards (BS 1377:1975) and calculated from equation 3-1:

$$Gs = \frac{Gl(m_2 - m_1)}{(m_4 - m_1) - (m_3 - m_2)} \qquad \text{3-1}$$

Where; Gl = Specific gravity of liquid (g/mL)

m_1 = mass of bottle (g)

m_2 = mass of bottle and dry soil material (g)

m_3 = mass of bottle, soil and liquid (g)

m_4 = mass of bottle and liquid (g)

The specific gravity was then used to calculate the volume of solids using the equation; $V_s = \frac{G_s}{M}$. Subsequently, the void ratio and porosity could be calculated from equations 3-2 and 3-3;

$$e = \frac{V_t - V_s}{V_s} \qquad \text{3-2}$$

$$n = \frac{e}{1 + e} \qquad \text{3-3}$$

The volume of cementation media pumped through a mould for one treatment cycle was equivalent to the pore volume of the substrate within the bio-brick mould. The two foam pieces were taken into consideration when calculating the pumped volume and it was assumed that its porosity was 75%. Thus, the volume of cementation media pumped through for one cycle was calculated to be 650 mL.

3.4.4 Bacteria and Inoculation

The bacteria was grown and harvested according to the method outlined in section 3.1. The cultures were mixed and diluted with fresh media to acquire an optical density (OD) of 1.0 ± 0.05. To inoculate the sand, the bacteria culture was mixed in an open beaker until the sand was saturated. The inoculated material was added into the bio-brick moulds with only the top lid open, allowing the liquid to drain through the bottom inlets. The moulds were then closed, and after four hours the first treatment of cementation media was pumped through each mould.

3.4.5   Setup

The experimental setup can be seen in Figure 3-5. The cementation media was stored in and pumped from a separate 25 L container and was connected by silicone pipes to a peristaltic pump. The pump was then connected to the three bio-brick moulds which were connected to three bottles that collected the effluent from the process.

Figure 3-5: A picture showing the laboratory setup for the bio-brick experiments.

The pump pumped the cementation media from the influent container into the bio-brick moulds for 250 seconds, at a flow rate of 0.0026 L/s, to fill the sands pore volume. The cementation media was retained in the bio-brick moulds for a set retention time to facilitate urea hydrolysis and allow for calcium carbonate precipitates to form.

3.4.6   Sampling

The efficiency of the treatments was monitored by sampling the influent cementation media's calcium, urea and ammonium concentrations once a day. The effluent calcium and ammonium concentrations were sampled twice a day at the end of the retention time while the solution was being pumped out of each mould. About 50 mL of the effluent was sampled at a time, and the measurements were taken immediately to avoid a change in pH because of a loss of ammonia due to volatilisation.

The difference in calcium concentrations was assumed to be the amount of precipitate formed place in the influent container. The effluent ammonium concentration monitored the uerolytic activity within the moulds. The amount of ammonium in the effluent indicated how much urea was degraded by urea hydrolysis.

The calcium and urea usage efficiencies were assumed to be the precipitated calcium molar concentration as a percentage of the molar calcium and urea concentrations in the influent respectively, and the equations used for this calculation can be found in Appendix B.

3.4.7 Compressive Strength Testing

The compressive strength was tested using three compressive tests shown in section 2.7. The bio-bricks produced were oriented on their side with the surfaces with the dimensions 222 mm X 75 mm in contact with the load and base of the compression testing machine or using brick laying terminology the bio-brick was placed in the shiner orientation. This orientation allowed for the smooth surfaces to resist the pressure more homogenously then the surface areas on the influent and effluent faces. The compressive strength was calculated using equation 2.6.1.

After the bio-bricks were removed from the moulds, a Proline Z100 (Zwick Roell, Ulm, Germany) machine was used to test the compressive strength. The size of the loading frame allowed the bio-brick to be tested in the shiner orientation and produced a stress-strain curve simultaneously. The load area was set to the dimensions of the side of a brick (222 mm by 73 mm), but in the case of deformed and broken bio-bricks, the compressive strength was calculated as a function of the area bearing the load.

3.4.8 Experimental Conditions

To achieve the aims of this book, experimental conditions were adjusted over several experiments. The conditions for each experiment are shown in Table 3-2. The first set of experiments, identified by the prefix A, aimed to test the feasibility of producing a bio-brick and establish an optimal starting influent calcium concentration. The second set of experiments, identified by prefix C, sought to establish what the maximum optimal influent calcium concentration was during the experiments. To achieve this, the calcium concentration was raised in a stepwise manner while the effects on the microbial activity within the bio-brick mould were assessed. Experiment R1 explored the effects a range of retention times had on the bio-brick system in order to establish an optimal retention time. Lastly, Experiments T1, T2 and T3 investigated the relationship between the number of treatments and the resultant compressive strength of the bio-bricks to determine the optimal number of treatments.

Table 3-2: Summarises the experimental conditions of all the bio-brick experiments conducted. The experiment names are delineated by the prefix letter according to their aim, either A, C, R or T. The numbers indicated the variation of the experiments conducted to achieve the aim. The starting influent calcium concentrations were increased between each experiment (A). The influent calcium concentration (C) was increased in a stepwise manner during each experiment. The retention time was varied throughout the experiment (R). The number of treatments increased for each experiment (T). The Table includes the starting influent, calcium concentration, the operating influent calcium concentration, the retention time, number of treatments before the mould was opened, number of repeats as well as the aim of each experiment.

Exp. Name	Influent calcium conc. [M]	Retention time [hours]	No. of treatments [-]	Aim
A1	0.045	4	48	To find the maximum starting influent calcium concentration
A2	0.09	4	48	
A3	0.11	4	48	
C1	0.09, 0.11, 0.13	4	48	To find the maximum influent calcium concentration the bio-brick process can withstand during the experiment using stepwise increases
C2	0.09, 0.1, 0.11, 0.12, 0.13	4	48	
R1	0.095	1, 2, 3, 4, 6 and 8	48	To investigate the effect of retention time on the bio-brick process
T1	0.095	2	38	To investigate the relationship between the number of treatments and the compressive strength of the bio-bricks
T2	0.095	2	58	
T3	0.095	2	68	

3.4.9 Feasibility of 'Growing' a Bio-brick and the Optimal Starting Influent Calcium Concentration

Experiments A1, A2 and A3 firstly sought to verify that the constructed moulds could produce a bio-brick using MICP and real urine. Secondly, the experiments explored how using different calcium starting influent concentrations affected the solidification, the amount of calcium carbonate precipitated in the moulds and the timescale of the process. Additionally, they sought to establish an optimal starting influent calcium concentration which was defined as the highest concentration the system could begin with from the first treatment, without causing detrimental effects on the microbial community. Each of these experiments maintained a retention time of 4 hours and kept the substrate in the mould for a total of 48 treatment cycles.

Each experiment had a different influent calcium concentration that remained constant throughout the experiment. Experiment A1, A2 and A3 had influent calcium concentrations of 0.05 M, 0.09 M and 0.11 M respectively. The effluents were monitored for the point at which the starting influent calcium concentration fails to induce MICP effectively, limiting microbial activity. The resultant optimal starting influent calcium concentration was recorded as the highest tested concentration before the concentration which reflected ineffective MICP.

3.4.10 Optimal Operating Influent Calcium Concentration

Experiments C1 and C2 were conducted to establish an optimal operating influent calcium concentration by raising the calcium concentration throughout the course of the experimental run. The optimal influent calcium concentration for the purposes of C1 and C2 is defined by the highest concentration that causes no detrimental effects on the microbial community. The effluents were monitored for the point where the influent calcium concentration begins to affect the microbial activity negatively. The resultant optimal influent calcium concentration was recorded as the highest tested concentration before the microbial activity was affected. The degree at which the influent calcium concentration was increased depended on what the effluent sample reflected. During both experiments, the influent calcium was increased after at least two samples reflected no signs the microbial community was adversely affected. Additionally, the influent calcium was lowered after 2 days if the effluent samples continually reflected adverse effects to the microbial community, this was done to observe if the system could re-stabilise.

From the results in experiments A1, A2 and A3 the maximum starting influent (0.09 M) was employed to start both experiments C1 and C2. Over the course of experiment C1, the influent calcium concentration was increased from 0.09 M on the 10th and 15th treatment to 0.11 M and to 0.13 M respectively and lowered to 0.09 M on the 35th treatment. Similarly, the progression of experiment C2 saw the rise in influent calcium concentrations from 0.09 M, to 0.1 M, to 0.11 M, to 0.12 M and to 0.13M on the 6th, 16th, 26th and 36th treatment respectively. Experiment C2 sought to find a new optimal concentration by increasing the influent calcium concentration between the optimal and the critical calcium concentrations established in C1, by smaller increments. During both experiments, the retention times were kept constant while the influent calcium concentrations were varied. After 48 treatments, the bio-bricks from both experiments were removed from their moulds.

3.4.11 Optimisation of Retention Time

As presented in Table 3-2, the retention time for the previous experiments was kept constant. However, the purpose of the experiment R1 was to investigate how the retention time affected the bacteria as measured by the ureolytic activity, the calcium carbonate precipitated and the pH in the effluent. Furthermore, the experiment aimed to find an optimal retention time that could reduce the overall time taken to produce a bio-brick but also sustain ureolytic activity.

The experiment began with a retention time of 4 hours, it was subsequently increased by 2 hours to 6 hours and then to 8 hours, to observe how a longer retention time affected the bacteria. It was then lowered back down to 4 and decreased by 1-hour increments to observe how reducing the retention time affects the system. The retention time was lowered after every second sample reading if the effluent did not reflect any adverse effects on the bacteria and it was raised back up if negative effects were detected to see if the system could re-stabilise. The retention time was decreased from 3 hours, to 2 hours and to 1 hour and after two samples were taken at 1 hour, it was raised back up to 2 hours.

The bio-bricks were kept in the mould for the same number of treatments as the previous experiment and removed from the moulds after 48 treatments. Experiments were conducted with a constant influent calcium concentration of 0.095 M.

3.4.12 The Relationship Between the Number of Treatments and Compressive Strength

Experiments R1, T1, T2 and T3 investigated the relationship between the number of treatments and the compressive strength of the bio-bricks produced. The experimental conditions are described in Table 3-2 and experiments T1, T2 and T3 implemented the optimal retention time of 2 hours found in experiment R1. The influent calcium concentrations emulated experiment R1 and began with a starting concentration of 0.095 M, which was kept constant for the entirety of each experiment. The number of treatments for each experiment varied, from 38, 48, 58 and 68. Once removed from the mould, all the bio-bricks were tested for their compressive strength.

3.5 Washing

When the bio-brick moulds were opened, a strong and unpleasant smell was experienced. This was a consequence of the ammonia being released from in between the pores of the bio-brick. It was observed that after a few days of leaving the bio-bricks to stand, the smell would dissipate as the ammonia escaped. Tests were executed to establish whether the smell could be removed by "washing" with tap water before the bio-brick moulds were opened.

The influent container was replaced with a container filled with tap water. A pore volume (650 mL) of water was pumped through the bio-brick moulds, and the ammonia in the effluent was measured according to the procedure outlined in section 3.3.3. The required volume was pumped through the bio-brick moulds every 15 minutes until the effluent reflected a negligible ammonium concentration.

3.6 Ionic Strength

The maximum influent calcium concentration observed from the bio-brick experiments – categorised by a peak in the urea hydrolysed – was found to be significantly lower than values obtained from literature (Okwadha & Li, 2010, Whiffin, Van Paassen & Harkes, 2007). It was postulated that the discrepancy was due to the higher ionic strength of the cementation media a solution and is calculated according to equation 3-4, where c is the concentration of the ion and z is the associated charge.

$$I = \frac{1}{2}\sum c_i z_i^2 \qquad\qquad 3\text{-}4$$

Several ions are present in the stabilised urine before the addition of the calcium chloride, whereas in the cementation media used in other studies, only urea, calcium chloride and a bacterial food source are present. This results in the ionic strength of the cementation media comprised of stabilised urine, being higher than that found in literature. An experiment was set up to observe how raising the ionic strength affected the urea hydrolysed. The range of ionic strengths used during the set of experiments reflected the ionic strengths used throughout the bio-brick experiments.

Table 3-3 indicates the ionic compositions of urine and the range of ionic strength in the cementation media used throughout the bio-brick experiment.

Table 3-3: Shows the ionic strength compositions and the resultant ion strength in fresh urine and the range of cementation media used in this dissertation's bio-brick experiments. The ionic strengths from the fresh urine and bio-brick experiment cementation media were calculated from the concentration composition found in (Randall et al., 2016) using PHREEQ, a thermodynamic programme.

		Ionic Strength [mol/L]	
		Cementation media used during the bio-brick experiments	
Species	**Fresh urine**	**Lower limit**	**Upper limit**
Acetate	0.106	0.106	0.106
Ammonia	0.03	0.03	0.03
P	0	0	0
Ca	0.029	0.045	0.131
Mg	0	0	0
Na	0.104	0.104	0.104
K	0.009	0.009	0.009
S(6)	0.025	0.025	0.025
Cl	0.125	0.159	0.194
N	0.001	0.001	0.001
OUTPUT	0.429	0.48	0.601

Five experiments were set up using solutions shown in Table 3-4 which corresponded to the varying ionic strengths found in Table 3-3. To mimic the urea concentration found in urine, synthetic urea was added to the solution at a concentration of 0.3 M. To isolate the effect of ionic strength only, NaCl was added to increase and acquire the desired ionic strength. Triplicate, autoclaved, Erlenmeyer flasks were filled with 250 mL of the specified solutions and then sealed with laboratory film to minimise ammonia volatilisation.

Table 3-4: Range of ionic strength concentrations for investigating the influence ionic strength has on the rate of urea hydrolysis.

Experiment No.	**Ionic strength mol/L**
1	0
2	0.2
3	0.4
4	0.6
5	0.8

A concentrated bacteria culture was grown according to section 4.1.2. The bacteria, *S. pasteurii*, was harvested at the end of the exponential phase at optical densities (OD) ranging from 1.05 to 1.13. The bacteria were then centrifuged, and this multiplied the OD by an average of 21 volume added to each Erlenmeyer flask was calculated to acquire an OD of 1 using a micropipette. The flasks were then placed on magnetic stirrers for 28 hours. The pH was monitored by placing a pH probe into one of the flasks for the duration of the experiment. All the flasks were sampled and tested for the ammonium concentration every hour for the first 3 hours, then every 3 hours for the next 9 hours, after which the flasks were sampled every 6 hours until the 28th hour of the experiment.

3.7 The Effects of pH and Calcium Concentration on MICP

Literature lacks substantial data on MICP operating at high pH's and high calcium concentrations, both of which are relevant to the operating conditions employed in this book. To deepen the understanding of the mechanics and relationships between the factors mentioned above, the following experiments were designed. More specifically the experiments were conducted to observe whether using a cementation media with an optimal pH of 9 for urea hydrolysis, instead of a pH of 11.2, results in an increase in ureolytic activity and greater calcium carbonate precipitation with higher calcium concentrations. The range of calcium concentrations were chosen from the results found in bio-brick experiments A2 and A3. The influent calcium concentration used in A2 (0.09 M) and A3 (0.11 M) showed a negligible effect on the MICP process and an inhibitive effect on the MICP process respectively. As a result, the concentrations used in A2 and A3 were used as the lower and upper limits for the experiment. Furthermore, a third calcium concentration between both the lower and upper limits of 0.1 M was tested.

The conditions of each experiment are shown in Table 3-4. Triplicate, autoclaved Erlenmeyer flasks (250 mL) were filled with 100 mL of cementation media prepared with the pH values and calcium concentrations stipulated. After the flasks were filled with the cementation media, they were sealed with laboratory film. Before conducting the experiments, a titration for a 100 mL of filtered stabilise urine was done to obtain a titration curve. A micropipette was used to add increments of 1 M of HCl, and at each increment, the pH was recorded. The titration was run in triplicates and informed the amount of HCl to add to each flask. To acquire the range of initial calcium concentrations in the flasks, the calcium chloride was added. However, before this, samples were taken to measure the calcium ion concentration, from which the amount of calcium chloride to add was calculated accordingly.

Table 3-5: The calcium concentration and pH that each flask of the experiment number.

Experiment No.	Ca^{2+} Concentration [M]	pH
1	0.11	11.2
2	0.11	9
3	0.1	11.2
4	0.1	9
5	0.09	11.2
6	0.09	9

A concentrated culture of bacteria was grown in accordance with the method described in 3.1.2. Using a glass pipette, the bacteria was added to each flask to acquire an OD of 1, mimicking the starting OD of the bio-brick experiment. The volume of bacteria culture added to the cementation media was calculated so that the flasks had an OD of 1. The bacteria, S. pasteurii, was harvested at the end of the exponential phase at optical densities (OD) ranging from 0.99 – 1.22. The bacteria were then centrifuged, and this multiplied the OD by an average of 17 times. To reach the optical density target, a volume of concentrated culture was added to each flask. The effect of this on the concentration and dilution was neglected.

After inoculating the flasks with the concentrated bacteria, the flasks were sampled over a period of 310 minutes for the dissolved calcium ion concentration and were taken at times of 0, 10, 30, 50, 80, 110 and 310 minutes. The pH was measured by placing a pH probe in each flask for the entirety of the test. Additionally, 100 mL of the cementation medias produced in Table 3-5 were poured into autoclaved Schott bottles. Samples were taken before the experiment began to measure the dissolved ammonium in the cementation media. Subsequently, concentrated bacteria was added to acquire an OD of 1, after which the Schott bottles were sealed tight to avoid ammonium being lost due to the volilitisation of ammonia gas. After 310 minutes, another sample was taken, and the dissolved ammonium was measured.

3.8 The Feasibility of Using a Cementation Media With a pH of 9

The results from section 3.7 indicate that a lower pH has a beneficial impact when used at higher concentrations of calcium ions thus instigating a further investigation into the effects of using a cementation media with an optimal pH of 9 (Maurer & Schwegler (2003) on spontaneous urea hydrolysis while also assessing the feasibility of using a cementation media with this pH in the bio-brick experiments.

Cementation media was prepared, and two 500 mL autoclaved Schott bottles were filled with 400 mL of the media. Using a pipette, 1 M of HCl was added to each bottle to lower the pH. The amount of HCl added to each flask was informed by the titration curve produced in 3.7. The pH of the first bottle was lowered to 11.2 and the second to 9. The Schott bottles were subsequently sealed and left to stand for eight days, during which samples were taken every 2nd day to measure the ammonium, urea, and calcium concentrations. All the sampling was conducted aseptically to ensure no contamination of the urease producing bacteria. For scientific purposes, the experiments were conducted in triplicates.

3.9   Alternative Nutrient Media

The use of standard media for bacteria growth is expensive and prohibits the field trials and real-world application of MICP technology. Van Drecht et al. (2009) stated that using ATCC® 1376 media to grow the bacteria would equate to 75% of the total operating costs of bio-brick production. Additionally, Rittmann et al. (2011) found that the cost of the growth mediums would increase proportionally with the scale of production. For this reason, it is critical to look for alternatives that stimulate and support bacterial growth and ureolytic activity for the bio-brick process to be economically feasible.

3.9.1   Alternative Nutrient Medias Chosen and Their Physiochemical Properties

S. pasteurii is a strain of *B. subtilis* which uses mainly ammonia as a nitrogen source in the absence of added growth factors (Mihelcic, Fry & Shaw, 2011). Work by Cuzman et al. (2015) identified several industrial effluents rich in protein that could be used as potential alternative nutrient medias (ANM). For the following investigation, three different industrial waste streams, similar to the effluents tested in Cuzman et al. (2015), that could be sourced locally in the Western Cape region, were chosen as ANMs. Brewhouse yeast (BW) was collected from a brewery (Drifter Brewing Co, Woodstock, Cape Town); Lactose Mother Liquor (LML) and waste whey were collected from a cheese processing plant (Klien River cheese, Standford, Overberg, Western Cape). BW was collected from the waste yeast dispelled from the fermentation tanks at the end of the fermentation stage of beer production. LML was collected from the industrial wastewater exiting the cheese production process, and Whey is the by-product of the acid-coagulated cheese, collected from the cheese production floor.

Once collected, the ANMs were analysed for their physio-chemical properties using the standard testing procedures outlined in the Water Quality Lab SOP manual (Dept of Civil Engineering, 2010). Total Kjeldahl Nitrogen (TKN), Organic Nitrogen (Org-N), Total Phosphorous (TP), Organic Phosphorous (Org-P) and Chemical Oxygen Demand (COD) were tested. TKN measures the total nitrogen concentration (inorganic and organic) in the sample and is directly related to the number of proteins present. The Chemical Oxygen Demand (COD) is used to quantify the number of organics in a sample. The pH and OD of each alternative nutrient media was

3.9.2 Bacteria Growth Testing

In order to delineate the growth stages and asses the feasibility of growth for *S. pasteurii* in the selected Alternative Nutrient Medias, a growth curve had to be produced. Separately, a growth curve for *S. pasteurii* grown in the conventional ATCC® 1376 media was determined for comparison.

Triplicate growth tests were conducted 3-4 days after the collection of the ANMs. In the interim period between collection and growth testing the ANMs were stored in a fridge at 4°C.

The bacteria culture was prepared and grown from a frozen stock with ATCC® 1376 according to the procedure outlined in 3.1. The bacteria culture was incubated at 32°C at 150 rpm overnight and grown to the late exponential/early stationary phase after which it was sub-inoculated into the ANMs.

Four 250 mL Erlenmeyer flasks were autoclaved for the growth experiments. Each Flask was filled with 50 mL of a different growth media; BW, LML, whey and ATCC® 1376. All the flasks were then inoculated with the same volume of the prepared bacteria culture in the ATCC® 1376 media to obtain a starting OD of 0.075. The flasks were incubated at 32°C at 150 rpm for 27 hours. The OD was measured every 2 hours and after 14 hours every 6 hours.

After 27 hours, the cultures were taken out of the incubator. A sterile loop wire was inserted into the culture and then aseptically streaked onto Christenson's Urea Agar (CUA) plates to detect any ureolytic activity. To make the CUA plates, 12 g of CUA was added to 500 mL of deionised water and then autoclaved. After being autoclaved the 20 g/L of urea was added aseptically while the mixture was still hot, ensuring a smooth, evenly mixed solution. The warm solution was then poured onto plates and kept in the fridge for later use. After the growth medias were plated, the plates were left for 48 hours. The plates indicate a pH change by turning red or pink as a result of urea degradation.

At the end of the growth experiments, the cultures were examined under a microscope (Olympus BX40, Tokyo, Japan) with a 100 times magnification.

All work with the bacteria was conducted in an aseptic manner which included autoclaving equipment and working under a fume hood.

3.9.3 The Rate of Calcium Carbonate Precipitation

The plates used in the previous experiment did not quantify the ureolytic activity but served as an indicator of the suitability of the alternative nutrient media. Using the CUA plates only served to distinguish between dead and viable bacteria cells and to detect ureolytic activity. It did not experiment. To test the viability of the cells, an experiment was set up to monitor the rate of calcium carbonate precipitation when a bacteria culture was inoculated into the bio-brick cementation media.

For the following experiment, stabilised, synthetic urine was used as a cementation media. The composition of the synthetic urine used by Henze & Randall's (2018) was used for this study and comprised of 0.3 M of urea, 3 g/L of nutrient broth (Nutrient Broth No 1, Sigma Aldrich, St Louis, United States) and 2 g/L of calcium hydroxide. The synthetic urine was then filtered through a filter paper with a diameter of 150 mm and a pore size of 1.2 µm (595 Schleicher & Schuell, Dassel, Germany) to remove excess precipitates. Using 1 M HCl, the pH was subsequently lowered to 11.2.

The stabilised, synthetic urine was poured into four 250 mL autoclaved Erlenmeyer flasks to a volume of 150 mL. Each flask was then inoculated with a different concentrated culture of bacteria grown using either; LML, whey, BW or ATCC® 1376 media. The concentrated bacteria cultures were prepared using the procedure outlined in 3.1.2. The volume of concentrated culture added to each flask was calculated to obtain an OD of 0.015. In order to reach the targeted OD in each solution, the volumes found in Table 4-4 of each of the centrifuged concentrated bacteria cultures (ATTC®1376, LML, whey, BW) were added to the flasks.

Table 3-6: A table to show the concentrated OD's after centrifuging and the volume of centrifuged culture added to the synthetic urine solutions

	ATCC® 1376	LML	Whey	BW
Concentrated OD	11.88	10.14	2.71	3.72
Volume of concentrated culture added (mL)	1.89	2.22	8.30	6.05

After the flasks had been inoculated with bacteria, they were sealed with aluminium foil. Samples and pH readings were taken at t = 0 hour, 2 hour, 4 hour, 6 hour and then subsequently left overnight and tested again at t = 24 hour, 28 hour and 32 hour. Each sample was measured with a syringe at 4-6 mL and filtered using a syringe filter. The samples were then diluted and tested for the dissolved calcium ion concentration.

After 32 hours the calcium carbonate precipitate was measured by filtering the contents of the flasks with a 90 mm diameter, micro-glass fibre filter papers with grade MG160 (Munktell Filtrek™, Bärenstein, Germany). The filter papers with the remaining solids were placed on crucibles and dried in an oven at 100°C overnight. To discard any organic material left on the filter papers, the crucibles were incinerated at 600°C. Before the filtration began, the crucibles

crucibles were weighed at each stage, holding; the wet solids and filter papers; the oven dried solids and filter papers and finally the incinerated solids. The weight of the incinerated solids was assumed to equal the measured calcium carbonate precipitated. Simultaneously a theoretical value for calcium carbonate precipitate was calculated from the change in concentration of dissolved calcium ions over the 32-hour testing period, using equation 3-5

$$m_{CaCO_3}[g] = \Delta C_{Ca^{2+}}\left[{}^g/_L\right] \times \frac{V_{cementation\ media}\ [L]}{\%\ Mass\ of\ Ca^{2+}\ in\ CaCO_3} \qquad 3\text{-}5$$

Chapter 4 Results

This chapter serves to report the findings from the experiments conducted according to the methodologies described from section 3.3 to 3.8. Section 4.1 reports on the findings pertaining to the feasibility and optimisation of the bio-brick production system. From these findings, further experiments which singled out and tested certain factors thought to be affecting the bio-brick system. These are described in sections 4.2 – 4.5. The factors include after treatment washing, ionic strength, pH and calcium concentration of the influent cementation media. Lastly, the chapter ends by reporting on the experiments conducted on alternative nutrient media for the bacteria strain used in this book. The findings observed in this chapter deal with the objectives outlined at the beginning of the book and specifically deal with objectives 2-5.

4.1 Bio-brick Experiments

4.1.1 Feasibility of 'Growing' a Bio-brick and Optimal Starting Influent Calcium Concentration

Figure 4-1 shows the initial steps of the MICP process under the microscope. Figure 4-1A shows *S. pasteurii grown* in ATTC° 1376 media at an OD of 1 and the rod-shaped bacteria cells at 2-4 µm. Figure 4-1B shows the stage where the loose sand and Greywacke aggregate is mixed with bacteria by showcasing one sand grain from the mix. Like Figure 4-1A, Figure 4-1B shows rod-shaped bacteria with a length of approximately 2-4 µm surrounding the sand grain, reflecting a sand aggregate mix inoculated with a culture of *S. pasteurii*. The bacteria cells are clustered around the surface of the sand grain to form a bio-film. Figure 4-1C shows the solids that form. The X-Ray Diffraction (XRD) analysis confirmed that the solid formed in the MICP process was calcium carbonate. The solids grown in the flask were found to contain mainly calcium carbonate in the crystal form of vaterite and quartz was also present in the solid (Appendix B.3). The sand grain in Figure 4-1D is from a specimen taken after the MICP process and in comparison, to Figure 4-1A, the grain is surrounded by crystals rather than only bacteria.

Figure 4-2 further shows the process on a macroscopic level. The single grain of Greywacke aggregate has clear, sharp edges before MICP and after MICP it is covered in bio-cemented solids. Figure 4-2 also shows the cementation of the loose Greywacke-sand mix into a solid after MICP with bio-cementitious material apparent.

Figure 4-3 shows the results for the bio-brick experiments A1, A2 and A3. The numeric values for each experiment can be found in Appendix B.2 (Table B-1, B-2, B-3). Figure 4-3A (experiment A1) are the results obtained for an average influent calcium concentration of 0.045 M. After the 12th treatment a calcium carbonate within the bio-brick mould. The urea usage efficiency ranged from 12.9% to 15.6%, and the pH of the effluent ranged from 9.15 to 9.34.

Figure 4-3B shows the results for an average influent calcium concentration of 0.09 M, which correlated with an increase in the urea usage efficiency range (24.9% to 41.1%) and a decrease in the effluent pH range from 9.15 - 9.34 to 8.72 – 8.83. The higher influent calcium concentration meant that more calcium carbonate was precipitating in the moulds for each treatment.

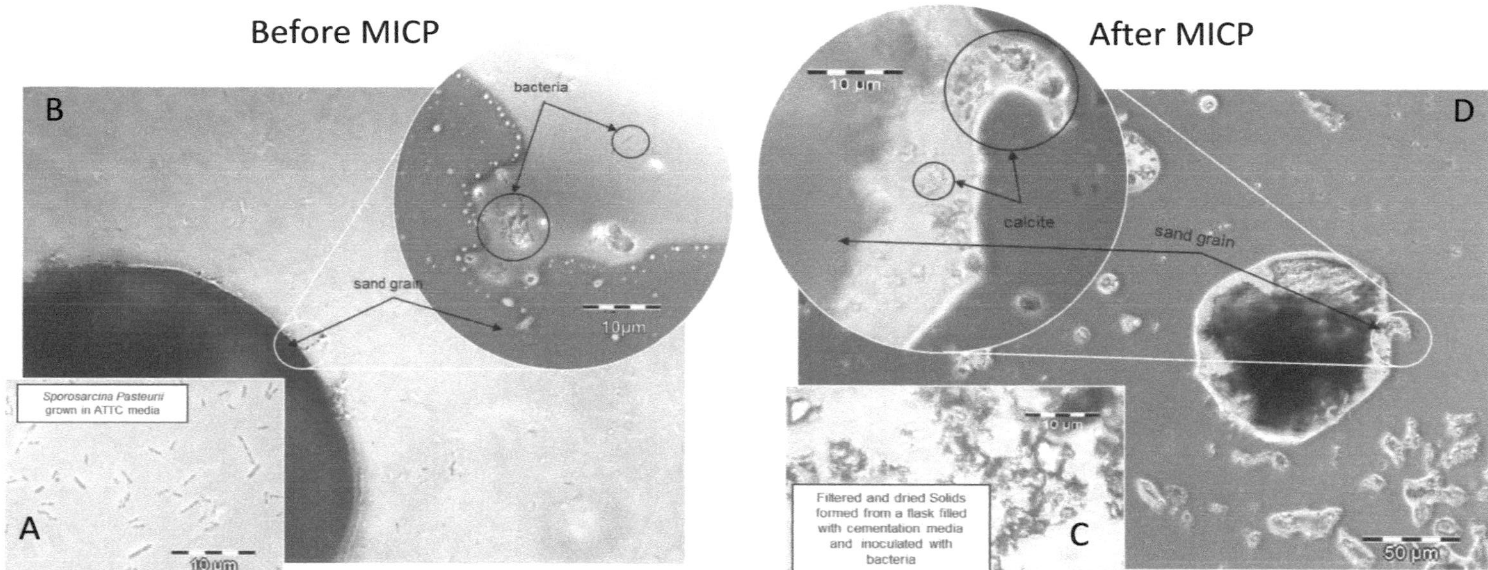

Figure 4-1: Microscopic images of various stages of the bio-brick cementation process. (A) *S. pasteurii* grown overnight according to section 3.1. (B) A sample taken from the inoculated (*S. pasteurii*) sand aggregate mix before the MICP process, showing a single sand grain. (C) The solids formed during a flask experiment mimicking the conditions of the bio-brick experiments. (D) A sample taken from the solid formed after MICP, specifically showing newly formed calcium carbonate crystals attached to a sand grain.

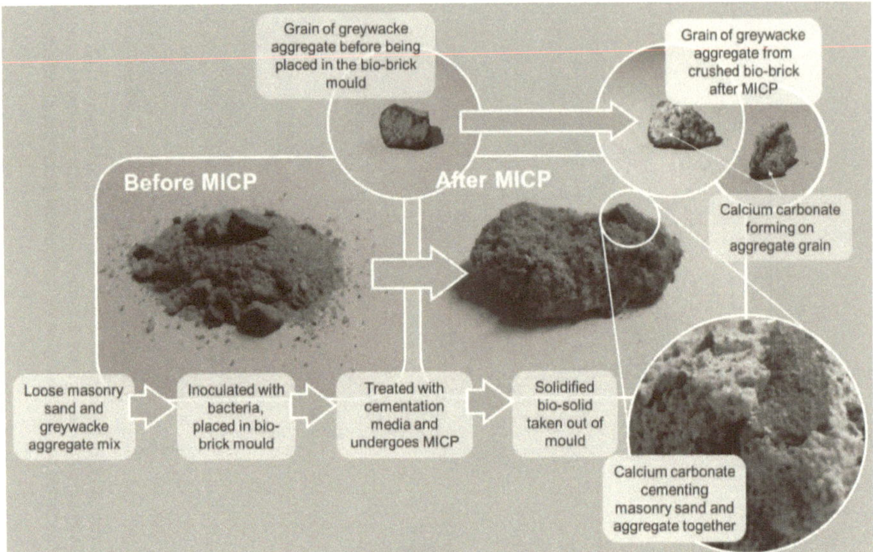

Figure 4-2: A schematic of macroscopic pictures taken of the loose sand-Greywacke mixture before MICP and a piece of the solid bio-brick formed after MICP. A close-up of the cemented bio-brick piece shows a bio-cementitious material cementing the sand-Greywacke mix together. Additionally, the schematic shows Greywacke aggregate grains before and after MICP. The bio-brick piece and Greywacke grain (after MICP) were both taken from a bio-brick that had been crushed to test for compressive strength. The schematic also details the process the loose sand has undergone in this research to become the solidified bio-brick.

The highest urea usage efficiency of 41.1% occurred at the peak influent calcium concentration of 0.1 M. The lowest influent calcium concentration of 0.087 M corresponded to the lowest urea efficiency of 27.7%. An outlier of this trend was observed on the 6th treatment where a lower urea efficiency of 24.9% corresponded with an influent concentration of 0.093 M.

Figure 4-3Cshows the results for an average influent calcium concentration of 0.11 M (experiment A3). The first sample reflected a deficient calcium and urea usage efficiency of 20.2% and 7.2% respectively. This further decreased until the 10th treatment to 2.1% and 0.8% where almost no urea was hydrolysed, and no calcium carbonate was deposited in the bio-brick mould.

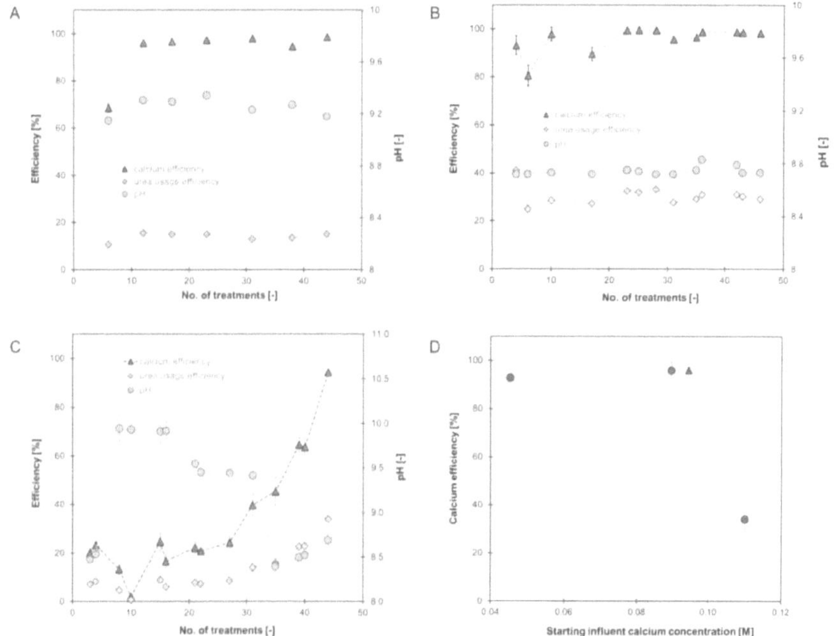

Figure 4-3: The calcium and urea efficiency for different influent calcium concentrations: (A) 0.045 M; (B) 0.09 M and (C) 0.11 M. The relationship between the influent calcium concentration and the average calcium efficiency is shown in (D). This figure includes the calcium efficiency for a varying influent calcium concentration obtained from experiment R1 (Δ).

This correlated with an increase in effluent pH from the 3rd to the 10th treatment at 8.5 and 9.9 respectively. At the 10th treatment, it appeared that no ureolytic activity was occurring. However, the bacteria appeared to revive and adapt to their environment as both the calcium and urea usage efficiency began to increase. The calcium and urea usage efficiencies continued to increase until the last treatment (48th) to 94.3% and 33.9% respectively. The effluent pH steadily decreased until the 31st treatment to measure 9.4 and then dropped by the 35th treatment to 8.4. Until the end of the experiment, the effluent pH rose from 8.4 to 8.7.

Figure 4-3D shows the average calcium efficiency corresponding to the varying influent calcium concentrations the experiments began with. The efficiencies exceeded 90% for influent concentrations 0.45 M, 0.09 M and 0.095 M (experiment R1) but decreased significantly to below 40% for an influent concentration of 0.11 M. However, comparing the average calcium removal efficiency in Figure 4-3D is not accurate, since the overall calcium removal is higher in mass basis in later treatments at 0.11 M initial concentration. It may indicate better system concentrations. The calcium usage efficiency for an influent calcium concentration of 0.11 M stayed below 25% for the first 27 treatments and then increased, reaching 94% after 44 treatments while the average calcium concentration was 33.9%. This limits the amount of calcium carbonate that can form and hence, the efficiency of the MICP process.

The theoretical cumulative percentage pore volume filled after 48 treatments for experiment A1, A2 and A3 were calculated as 11.7%, 27.8% and 5.8% respectively. After 48 hours, the moulds were opened, and Figure 4-4 depicts the range in solidification. The degree of solidification was positively correlated with the theoretical cumulative percentage pore volume filled. Figure 4-4 shows no signs of solidification for experiment A3, partial solidification for experiment A1 and a solidified bio-brick for experiment A2. Further notes on changes in the methodology from A1-A3 is described in Appendix B.3.

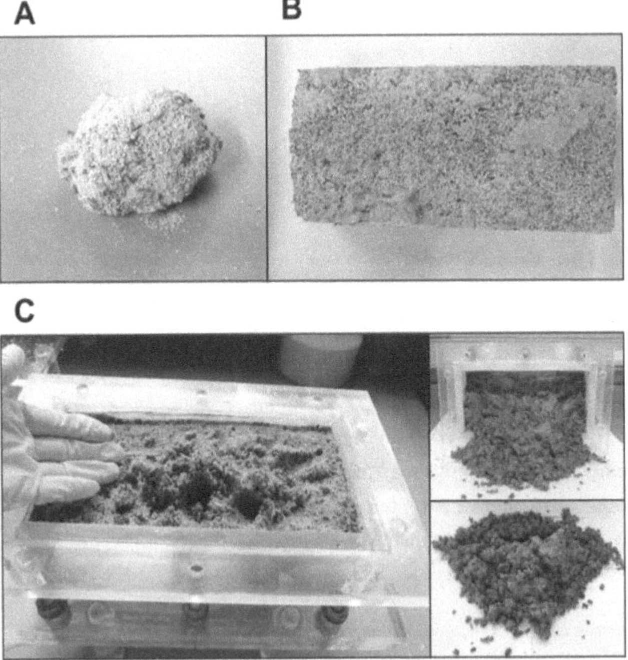

Figure 4-4: Photographs of the partial solidification of the bio-brick removed from the mould for experiment A1 (A). A solid bio-brick formed after 48 treatments in experiment A2 (B). No solidification occurred in experiment A3 after 48 treatments (C).

4.1.2   Optimal Operating Influent Calcium Concentration

The results for experiment C1 are shown in Figure 4-5, and the numeric values found in Appendix B-2 (Table B-4). The experiment began with a starting influent calcium concentration of 0.09 M, and by the 5th treatment, the bacteria were precipitating all the calcium fed into the mould, reflected by a high calcium efficiency (96%). The high calcium efficiency remained until the 14th treatment (96% to 100%). Additionally, during this period, the influent calcium was increased twice, once on the 5th treatment from 0.09 M to 0.1 M and secondly on the 10th treatment from 0.1 M to 0.11 M. Both these increases corresponded to increases in the urea hydrolysed. As the influent concentration was raised from 0.09 M to 0.1 M, the urea hydrolysed increased from 0.11 M to 0.18 M. The second increase from 0.1 M to 0.11 M corresponded to an increase in precipitated calcium and urea hydrolysed from 0.09 M to 0.11 M and 0.15 M to 0.18 M respectively. In parallel, the urea usage efficiency also increased from 22% to 43%.

On the 17th treatment, the influent calcium concentration was further increased to 0.13 M. At first, the calcium efficiency remained high at 99% but subsequently began to decrease to 85% until the 34th treatment. The same pattern was observed in the urea usage efficiency decreasing from 43% to 37%. The urea hydrolysed did not follow the same trend, and a decrease was immediately observed when the influent calcium was increased to 0.13 M. The concentration of urea hydrolysed decreased to 0.16 M and continued to fall to 0.04 M over the next 17 treatments.

To combat the decline in ureolytic activity, after the 36th treatment the influent calcium was lowered back down to 0.09 M. Subsequently, 95% of the calcium from the influent precipitated in the mould, increasing to 99% after 5 more treatments. The urea hydrolysed increased to 0.194 M after 5 treatments of the lowered influent calcium concentration, showing a revival in the microbial community.

From the beginning of the experiment to the 5th experiment, the pH in the effluent steadily decreased from 9.0 to 8.5 but increased to 8.7 by the 10th treatment. As the calcium concentration increased to 0.13 M, the effluent pH dropped significantly to 8.1 but gradually increased to 8.5 over a period of 17 treatments. After the calcium concentration decreased to 0.095 M, the effluent pH continued to increase until the end of the experiment to reach 8.9.

After 48 treatments, 29% of the theoretical cumulative percentage pore volume of the bio-bricks had been filled with precipitated calcium carbonate. All three moulds were opened after 48 treatment cycles. In two of the moulds, the contents had solidified and could be handled as one solid piece. The contents of the third mould had not solidified due to a blocked influent nozzle. The solidified bio-bricks were tested for compressive strength, and Figure 4-7 shows a picture of them after testing. The surfaces facing the influent and effluent lids were both rough and uneven. The compressive strength results are further discussed in section 4.1.4.

Figure 4-6 graphs the results for experiment C2. The data used for this figure can be found in Appendix B-2 (Table B-5). The starting influent calcium concentration of 0.09 M was used, and by the 3rd treatment, 100% of the influent calcium was utilised (calcium efficiency). A continually high calcium efficiency (>99%) was observed from the 3rd to the 26th treatment. Additionally, during this period, the influent calcium was increased twice, once on the 8th treatment from 0.09 M to 0.10 M and secondly on the 17th treatment from 0.10 M to 0.11 M. Both these increases corresponded to a drop in the urea hydrolysed followed by a subsequent gradual increase. As the influent calcium concentration was raised from 0.09 M to 0.10 M, the urea hydrolysed decreased from 0.15 M to 0.13 M. After 6 treatments it increased back up to 0.15 M. At the increased influent calcium concentration of 0.11 M, the urea hydrolysed decreased from 0.15 M to 0.138 M, increasing up to 0.189 M after 2 treatments on the 28th treatment.

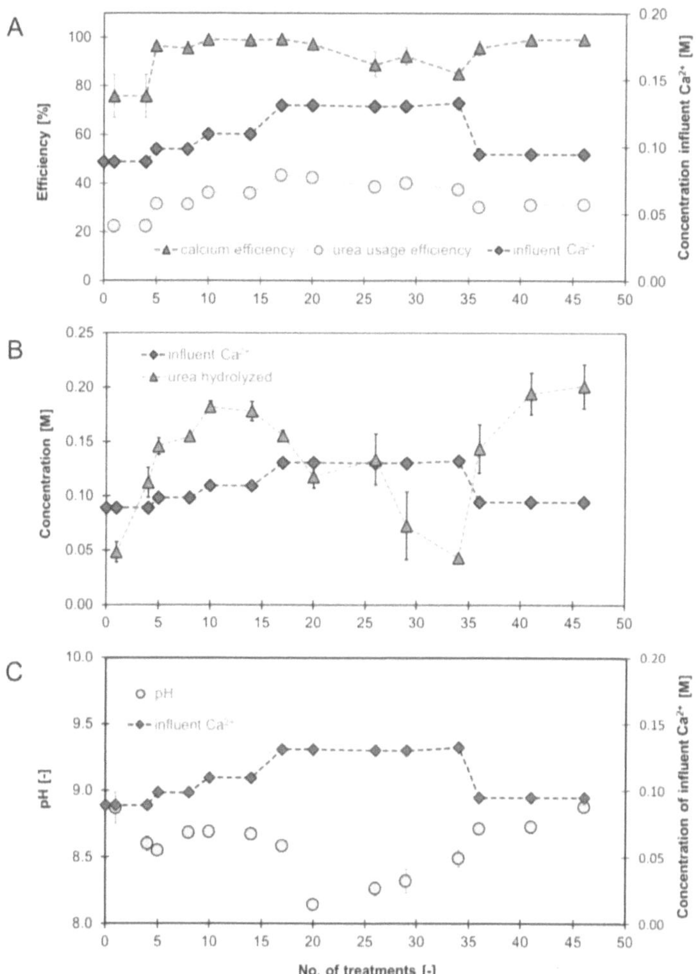

Figure 4-5: The influent calcium concentration, calcium and urea efficiency, urea hydrolysed and pH for experiment C1 are given for the range of treatments. The experiments were conducted in triplicate, and the error bars give the standard deviation for three bio-bricks. The calcium and urea usage efficiency over the 48 treatments and the influent calcium concentration fed into the moulds over the 48 treatments are also shown (A). The concentration of urea hydrolysed and influent calcium concentration over the 48 treatments (B). The pH of the effluent and the influents calcium concentration over the 48 treatments (C).

On the 28th treatment, the influent calcium was raised to 0.12 M, and after 4 treatments, the calcium efficiency decreased to 92%. However, after the 38th treatment, the calcium efficiency increased back up to 98%. The influent concentration increase caused the urea hydrolysed to decrease to 0.089 M but mirroring the calcium efficiency it increased back up to 0.118 M by the 38th treatment. On the 40th treatment, the influent calcium was increased to 0.13 M, and after 4 treatments the urea hydrolysed decreased to 0.80 M, and the margin between influent and precipitated calcium began to widen as the calcium efficiency dropped from 92% to 88%.

From the 1st treatment to the 5th treatment, the pH in the effluent steadily decreased from 8.97 to 8.54 but increased to 8.69 by the 10th treatment. As the calcium concentration increased to 0.13 M, the effluent pH dropped significantly to 8.14 but gradually increased to 8.5 over a period of 17 treatments. After the calcium concentration decreased to 0.10 M, the effluent pH continued to rise to a value of 8.88 on the 46th treatment. Troughs in the effluent pH graph correspond to points where the calcium concentration was increased. When the calcium concentration is increased the effluent pH decreases. However, after each decrease, the pH manages to stabilise.

All three moulds were opened after 48 treatment cycles when theoretically 31% of the pore volume of the bio-brick was filled with precipitated calcium carbonate. All three of the moulds produced solid bio-bricks which could be handled. Figure 4-7 shows all three bricks produced from experiment C2 before being crushed. One of the bio-bricks broke in half and was used for coring tests. Coring the bricks allows them to undergo extensive material property testing. Water absorbance and porosity can only be measured using cylindrical cored samples. However, the broken bio-brick pieces disintegrated during coring, and as a result, only compressive strength could be tested. The one half of the broken bio-brick was measured for compressive strength, but the value was adjusted according to the smaller surface area being tested. Only two of the bio-bricks were tested, and a third was kept for display purposes. The compression test results are elaborated on in section 4.1.4.

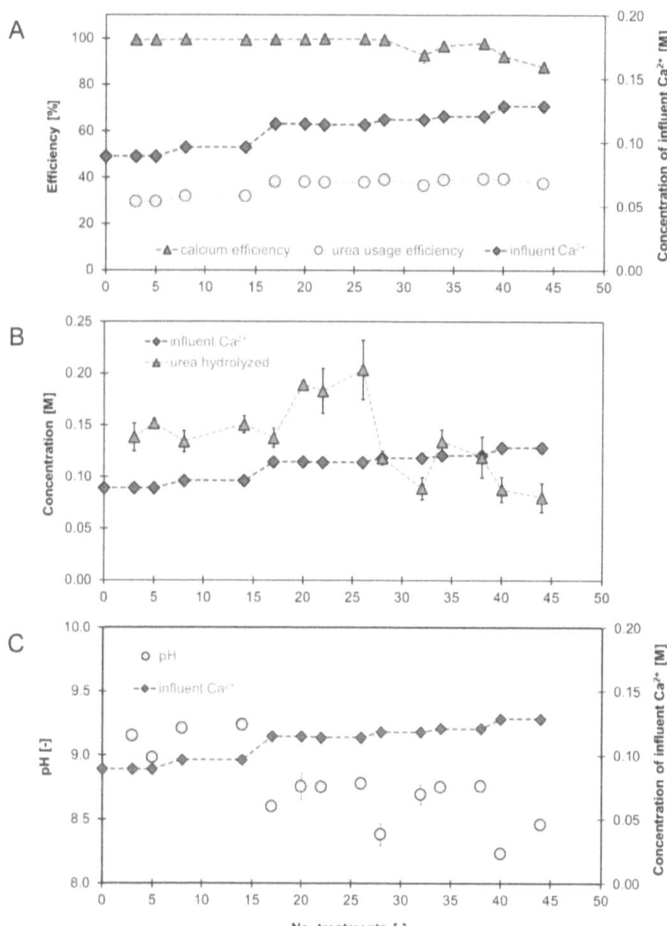

Figure 4-6: The influent calcium concentration, calcium and urea efficiency, urea hydrolysed and pH for experiment C2 are given for the range of treatments. The experiments were conducted in a triplicate, and the error bars give the standard deviation for three bio-bricks. The calcium and urea usage efficiency over the 48 treatments and the influent calcium concentration fed into the moulds over the 48 treatments (A). The concentration of urea hydrolysed and influent calcium concentration over the 48 treatments (B). The pH of the effluent and the influents calcium concentration over the 48 treatments (C).

Figure 4-7: Photographs of the two solidified bio-bricks produced in C1 after the compression test (A). The three, solid bio-bricks formed after 48 treatments in experiment C2 before compressive strength testing, the third was used for display purposes (B).

Figure 4-8 combines the data from both experiment C1 and C2 to show the correlation between the specific influent calcium concentrations and calcium efficiency, urea usage efficiency and the urea hydrolysed. At 0.09 M, the average calcium efficiency is 88%. This average, however, consists of samples taken in the first 5 treatments of the bio-brick experiments when the microbial community is still adapting to their environment thus reflecting lower calcium efficiencies, and the error amongst the values is 17%. The calcium efficiency increased to 96% at 0.01 M. The calcium efficiencies associated with the influent calcium concentrations of 0.115 M, 0.11 M and 0.095 M are all within the range 98% to 100%. The calcium efficiency drops to 96% for the influent calcium concentration of 0.12 M and to 92% for the influent concentration of 0.13 M.

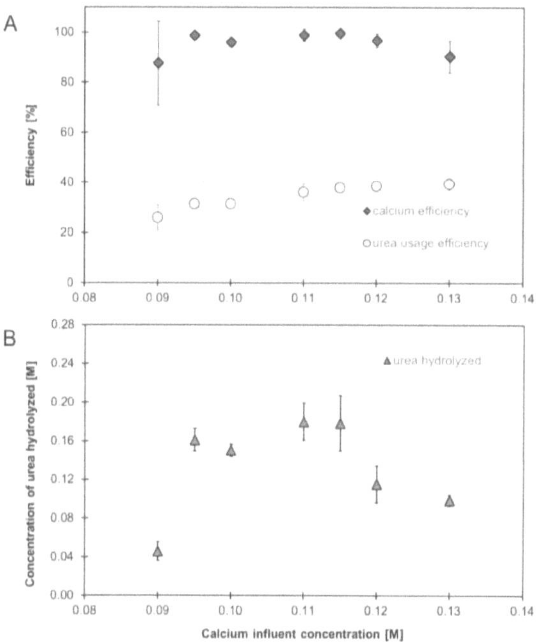

Figure 4-8: Average of measured variables for different influent calcium concentrations fed into the moulds during experiment C1 and C2. The average calcium and urea usage efficiency of each influent calcium concentration (A). The average urea hydrolysed corresponding to each influent calcium concentration (B).

The urea usage efficiency increased as the calcium concentration increased. It reached a maximum of 40% at the maximum calcium concentration of 0.13 M fed into the moulds. The amount of urea hydrolysed increased as the calcium concentration increased from 0.09 M to 0.115 M. At a calcium concentration of 0.12 M the amount of urea hydrolysed decreased from an average of 0.17 M to 0.12 M. The hydrolysed urea further decreased to 0.09 M for the influent calcium concentration of 0.13 M.

The effluent pH's were plotted against their corresponding influent calcium concentration in Figure 4-9. The scatter plot shows a general trend of higher influent calcium concentrations resulting in lower effluent pH's, and lower influent calcium concentrations achieving higher effluent pH's.

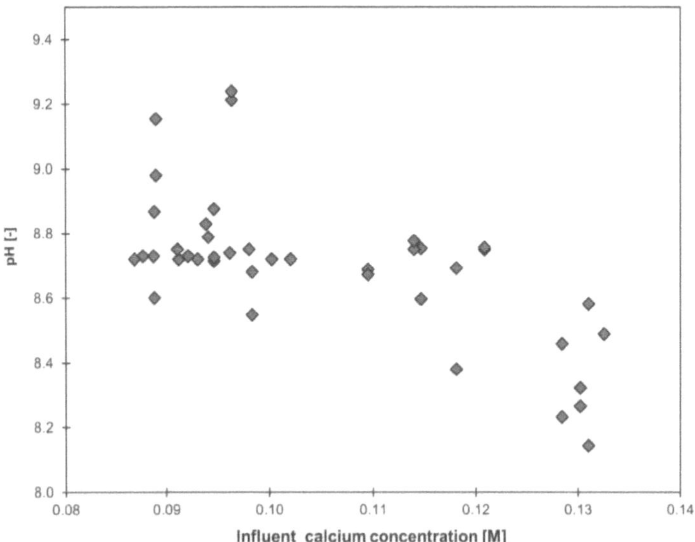

Figure 4-9: A scatter plot of the effluent pH's leaving the bio-brick moulds for various influent calcium concentrations.

4.1.3 Optimisation of Retention Time

The data for experiment R1 can be found in Appendix B.2 (Table B-6). As discussed in section 4.1.1, experiment A1, A2 and A3 established the maximum starting influent calcium concentration of 0.09 M. Experiment R1 used a new starting calcium concentration of 0.095 M which was below the limits found in the previous experiment but above the established maximum. By the 6th treatment, 98% of the calcium concentration fed into the mould was used for calcium carbonate precipitation (calcium efficiency). Figure 4-10A shows the calcium and urea usage efficiency for different retention times (experiment R1). The average calcium efficiency for retention times 2 to 8 hours all remained high (>97%). After lowering the retention time to 1 hour, the average calcium efficiency dropped to 86%. When the retention time was lowered from 2 to 1 hour, the calcium efficiency experienced a continual drop, as seen in Appendix B (Table B-6), and the average was calculated over this drop.

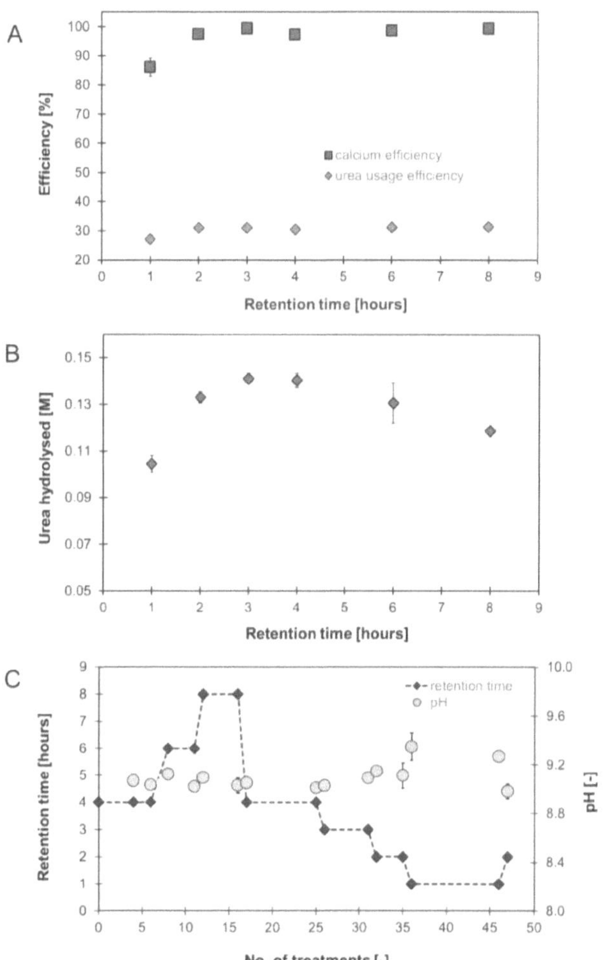

Figure 4-10:The average calcium, urea efficiency and urea hydrolysed in experiment R1 are given for a range of retention times and the pH in experiment R1 are given for a range of treatments. The experiment was conducted in triplicate, and the error bars give the standard deviation for three bio-bricks. The average calcium and urea usage efficiency for the range of set retention times from 1 hour to 8 hours (A). The amount of urea hydrolysed for the range of set retention times from 1 hour to 8 hours (B). The pH of the effluents and the retention times over the 48 treatments (C).

The calcium efficiency was 16.8% lower for a retention time of 1 hour as compared to the other retention times. Additionally, the urea usage efficiency also experienced a drop at a retention time of 1 hour. With retention times 2, 3, 4, 6 and 8 hour, the average urea usage efficiency ranged from 30.5% to 31.3%. However, when lowered to 1 hour, the average urea usage efficiency dropped to 27.2%. Again, this average was calculated over a continual drop in the urea usage efficiency for 20 treatments. Overall the efficiency decreased by 5.3%.

Figure 4-10B shows the average concentration of urea hydrolysed for each retention time, which is an indicator of the ureolytic activity. Retention times 3 and 4 hours experienced a peak in urea hydrolysis at 0.14 M. These dropped for both retention times 2 and 6 hour to 0.13 M. It dropped again at 8 hours to 0.12 M and the lowest urea hydrolysed occurred at a retention time of 1 hour (0.10 M). Again, the average for each retention time was calculated over a continual drop in the urea hydrolysed over 20 treatments. The concentration decreased by 0.037 M at a retention time of 1 hr after which the retention time was increased to 2 hours.

Figure 4-10C shows the pH over the course of treatments for experiment R1. The pH of the increased retention times remained in the range of 9 to 9.1. This data established that there were no significant detrimental effects on the bio-brick system if the retention time was set to 8 hours. It was subsequently decided to test the lower limits of the retention time, and, on the 17th treatment, the retention time was lowered back to 4 hours. Thereafter, it was lowered by one hour every day. For retention times 4, 3 and 2 hours, the pH in the effluent remained in the same range as the increased retention times (9 to 9.15). However, when the retention time was reduced to 1 hour, the pH in the effluent increased to 9.35.

After 48 treatments, the moulds were opened, and 3 solid bio-bricks were removed, all of which can be seen in Figure 4-11B. After 48 treatments, 24.6% (calculated) of the pore volume had been filled. The compressive strength results are further discussed in section 4.1.4.

Notes on observations and small adjustments to the methodology from experiment C1-C2 are described in Appendix B3.

4.1.4 *The Relationship Between the Number of Treatments and Compressive Strength*

The data derived from experiments T1, T2 and T3 can be found in Appendix B.2 (Table B-7, B-8, B-9). These experiments used a retention time of 2 hours. After 38 treatments, experiment T1 was stopped, and three bio-bricks were taken out of their moulds as shown in Figure 4-11A. All three bio-bricks were not fully formed with crumbling edge, and one of the bio-bricks broke into two pieces. Similarly, experiment T2 was stopped after 58 treatments, and the three bio-bricks solidification than that of the bio-brick obtained in experiment T1. However, the edges did crumble when touched.

Figure 4-11C shows the results for experiment T3 where three bio-bricks, opened after 68 treatments, after undergoing compressive strength testing. A quarter of one bio-bricks broke off, and the two sections were measured for their compressibility separately according to their dimensions. Another brick had half its edge removed (due to crumbling) when it was removed from the mould.

Figure 4-12 shows the average compressibility strength of the bio-bricks produced experiment A2, C1, C2, R1, T1, T2 and T3. The derived data can be found in Appendix B.4. As shown in Figure 4-12, the greater number of treatments, the higher the compressive strength. The compressive strength was also greater when the influent calcium concentration was increased in a stepwise manner. This is because the bacteria could acclimatise better and precipitate a more substantial amount of calcium carbonate, as seen in Figure 4-5 and Figure 4-6. Likewise, the number of days treated corresponded to an increase in compressibility. For both graphs, the plot points with a constant influent calcium concentration and ranging number of treatments (identified by the blue diamonds).

Table 4-1 gives a summary of the compressive strengths obtained from literature as well as the compressive strength obtained in this research. The maximum compressive strength obtained in this study was 2.7 MPa. It exceeds all the compressive strengths of the bio-solids produced by MICP using synthetic urea and *S. pasteurii*. Additionally, the max strength of 2.33 MPa from Bernardi et al., (2014) was created from 84 treatments over 28 days. The strongest bio-brick produced in this study was produced using approximately half the number of treatments in approximately a third of the time. The South African National Standards (SANS) compressive strength standards for face brick range from 9 to 12.5 MPa and non-facing brick range from 3-10.5 MPa (SANS 227, 2007). The maximum compressive strength obtained in C1 at 2.7 MPa is slightly below the minimum standard for non-facing bricks.

Figure 4-11: Bio-bricks produced from experiments T1 (A), R1 (B), T2 (C) and T3 (D). The number of treatments increased from 38, 48, 58 and 68, respectively.

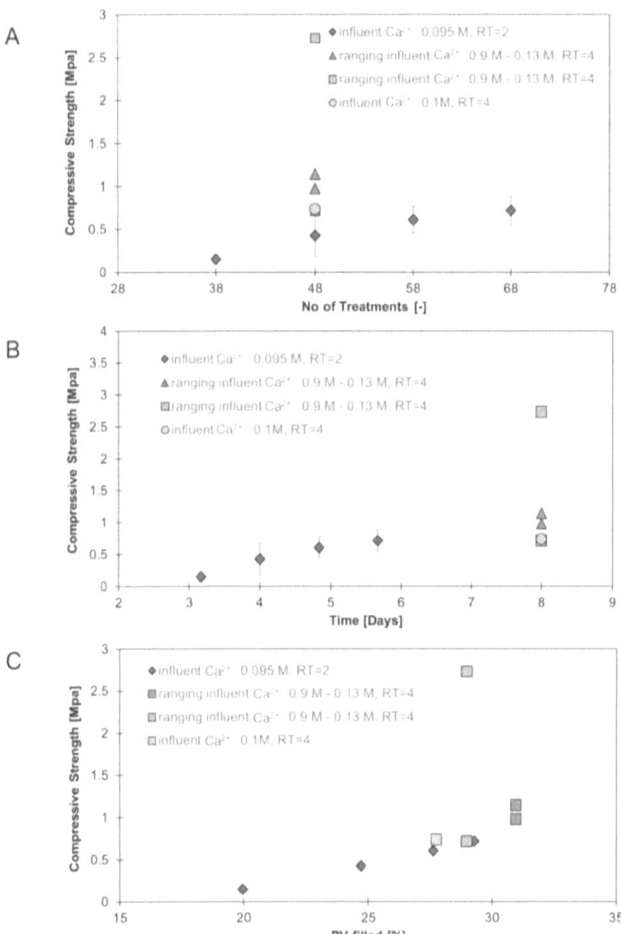

Figure 4-12: The compressive strengths achieved by the bio-bricks produced experiments R1, T1, T2 and T3 for a range of number of treatments, days treated, and percentage pore volume (PV) filled. The compressive strength corresponding to each the no. of treatments (A). The compressive strength corresponding to the total time treated in days (B). The compressive strength corresponding to Pore Volume (PV) filled as a percentage at the end of a treatment cycle (C). The bio-bricks produced in experiments R1, T1, T2 and T3 were produced with the same constant influent calcium concentration of 0.095 M (♦). The bio-bricks in experiment C2 were produced with a range of influent calcium concentrations increased in a stepwise manner (0.09, 0.1, 0.11, 0.12, 0.13 M) (▲). The bio-bricks in experiment C2 were produced with a range of influent calcium concentrations increased in a stepwise manner (0.09, 0.11, 0.13 M) (■). The bio-bricks produced in Exp. A2 were with a constant influent calcium concentration of 0.09 M (○).

Table 4-1: Table of different building materials and their compressive strengths.

Sample type	Bacteria	Compressive strength [MPa]	References
40% lime	-	0.12 min 0.93 max	(Bernardi et al., 2014)
20% cement	-	2.00	(Bernardi et al., 2014)
Solid clay brick	-	13.3	(Dehghan et al., 2018)
Bio-solid made using MICP	*Bacillus megaterium* (ATCC 14581)	0.76	(Lee, Ng and Tanaka, 2013)
Bio-solid made using synthetic urea	*S. pasteurii* (ATCC 11859)	0.12 (21 treatments) 0.93 (42 treatments) 2.33 (84 treatments)	(Bernardi et al., 2014)
Bio-solid made using synthetic urea	*S. pasteurii* (ATCC 11859)	0.29 min 0.87 max	(Henze & Randall, 2018)
Bio-bricks made using human urine	*S. pasteurii* (ATCC 11859)	2.7 max (exp C2) 0.15 min (exp T1)	this study

4.2 Washing

Figure 4-13 shows the concentration change in ammonium ions after each pore volume wash with deionised water. The ammonium concentration in the effluent from the last treatment of the bio-brick experiment, before the first wash was 0.18 M and by the third wash the concentration of ammonium was almost all depleted, with a value of 0.002 M. By the 4th and 5th wash, a negligible amount of ammonium ions remained. When the bio-brick moulds were subsequently opened, it was observed that no strong ammonia smell remained.

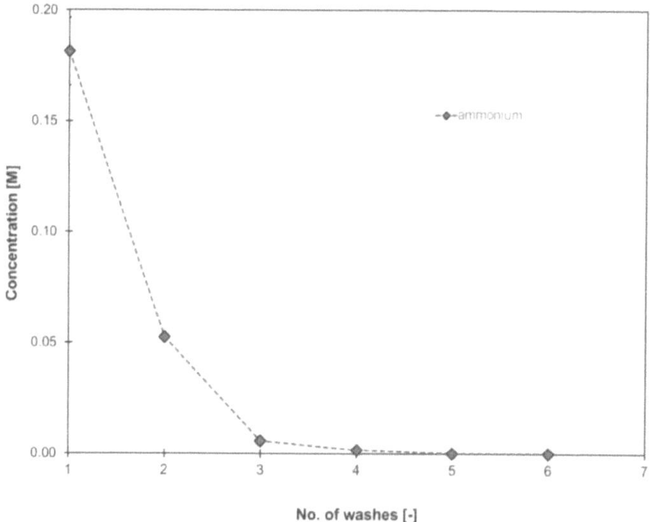

Figure 4-13: The change in ammonium ion concentration in the effluent as a pore volume of water (number of washes) are pumped through the bio-brick moulds.

4.3 Ionic Strength

Figure 4-14 shows the course of pH values (A) and ammonium concentrations (B) of synthetic urea solutions with different ionic strengths after being inoculated with a concentrated culture of bacteria. A trend is observed between the ionic strength and the amount of urea hydrolysed. The greater the ionic strength, the smaller the ammonium concentration found in the solutions. The maximum ammonium concentration observed was 0.2 M found in the solution with the lowest ionic strength. A relationship is also evident between the ionic strength and the pH of the media. The lower the ionic strength, the quicker the solutions reached a higher pH, plateauing at around 9.3. The highest ionic strength (0.8 M) did not reach a pH of 9.3 but plateaued at a later stage (after 24 hours) and at a lower pH of 8.7.

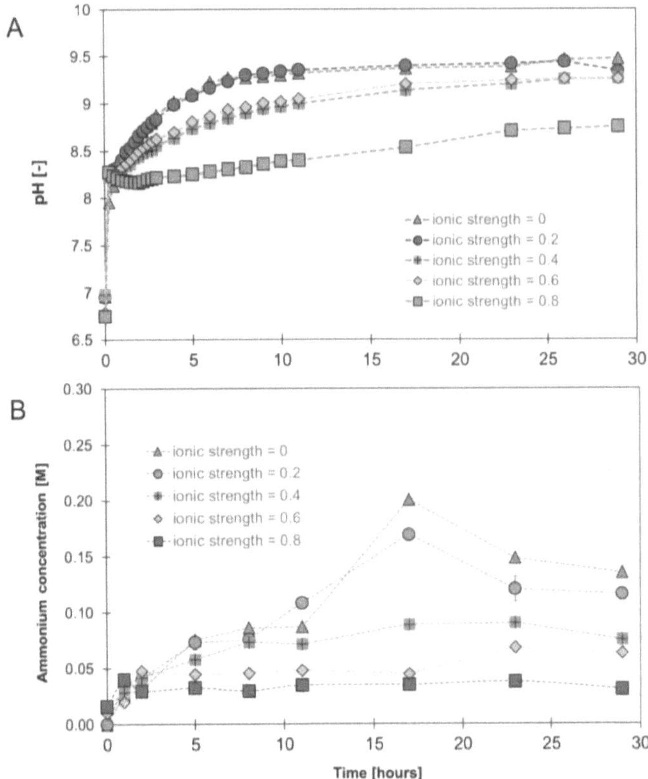

Figure 4-14: (A) The change in pH for different ionic strengths as a function of time (B) The change in ammonium concentration for synthetic urea solutions at a constant concentration of 0.3 M with varying ionic strengths (different concentrations of sodium chloride).

Figure 4-15 shows the rate of urea hydrolysis as a function of ionic strength based on work published by Kistiakowsky & Shaw (1953). It also shows the range of ionic strengths for the cementation media (0.48 − 0.60 M) used in the bio-brick experiments as well as the ionic strength of at typical fresh urine sample (395.2 mM).

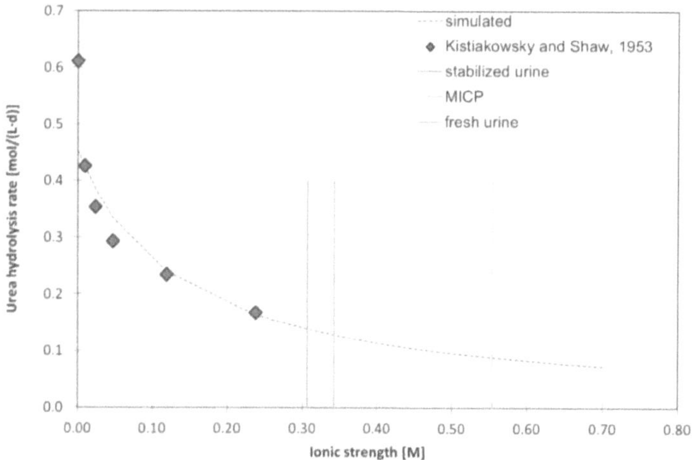

Figure 4-15: The rate of urea hydrolysis as a function of ionic strength together with the typical ionic strengths of fresh and stabilised urine as well as the range of ionic strengths for the bio-brick (A). The data for fresh and stabilised urine was taken from (Randall et al. 2016), and the ionic strength was determined from PHREEQC using the same method described (Randall et al. 2016). The simulated urea hydrolysis rate was calculated assuming non-competitive inhibition and the equation derived by (Brison 2016) and the urea hydrolysis rates were obtained from (Kistiakowsky and Shaw 1953).

4.4 The Effect of Influent PH and Calcium Concentration on MICP

Figure 4-16A shows the course of the average precipitated calcium in each flask of solution (cementation media) with different initial calcium concentrations and pH values. Three solutions had the same initial pH (11.2) but differed in their initial calcium concentrations (0.09, 0.1, 0.11 M). For these solutions, the graph shows that at the beginning of the experiment the rate of initial calcium carbonate precipitation was highest for the solution with an initial calcium concentration of 0.1 M. This was greater than both the concentrations 0.11 M and 0.09 M. The rate, however, decreased significantly in the 0.1 M solution flask, and after 80 minutes the precipitated calcium was 0.057 M and further sampling revealed more precipitation until the end of the experiment which reflected a total calcium precipitation of 0.071 M. Similarly, the flask with the lowest influent calcium concentration (0.09 M) also showed a decrease in the rate of calcium carbonate precipitation after 80 minutes, however, the rate was still greater than the rate for the 0.1 M initial calcium concentration solution and the calcium carbonate precipitation continued in a linear fashion until, at the end of the testing period, 100% of the initial calcium concentration of 0.09 M was precipitated.

After 30 minutes, the solution with the initial calcium concentration of 0.11 M reflected a precipitated (0.008 M) until the end of the testing period. The final amounts of calcium precipitated were equivalent to 27% (0.11 M), 71% (0.1 M) and 100% (0.09 M) of the initial calcium concentration, showing that increasing the initial calcium concentration effectively lowered the amount of calcium carbonate precipitated.

For the solution with an initial pH of 11.2, in the first 5 minutes, the pH of all three solutions dropped dramatically, however, the solution with the highest initial calcium concentration dropped the least to 10.2 and remained at this pH until the end of the duration of the experiment. The other two solutions reflected a greater drop in pH, 9.3 for the 0.09 M solution and 8.4 for the 0.1 M solution. The pH of the first solution (0.09 M) was slowly declining to the end of the experiment where it reached a final pH of 8.85, while the second solution (0.1 M) slowly increased to a pH of 8.6.

Three other solutions with a lower pH of 9.25 had the same variation of initial calcium concentrations (0.09, 0.1, 0.11 M). For these solutions, at 80 minutes the rate of calcium carbonate precipitation was greatest for the solution with an initial calcium concentration of 0.09 M, the rate decreased after this point, but remained at a constant rate for the duration of the experiment. The solution with an initial calcium concentration of 0.1 M maintained a relatively constant precipitation rate for the duration of the experiment. However, for the solution with an initial concentration of 0.11 M, the rate decreased significantly after 110 minutes after which only a small amount precipitated until the end of the experiment (0.009 M).

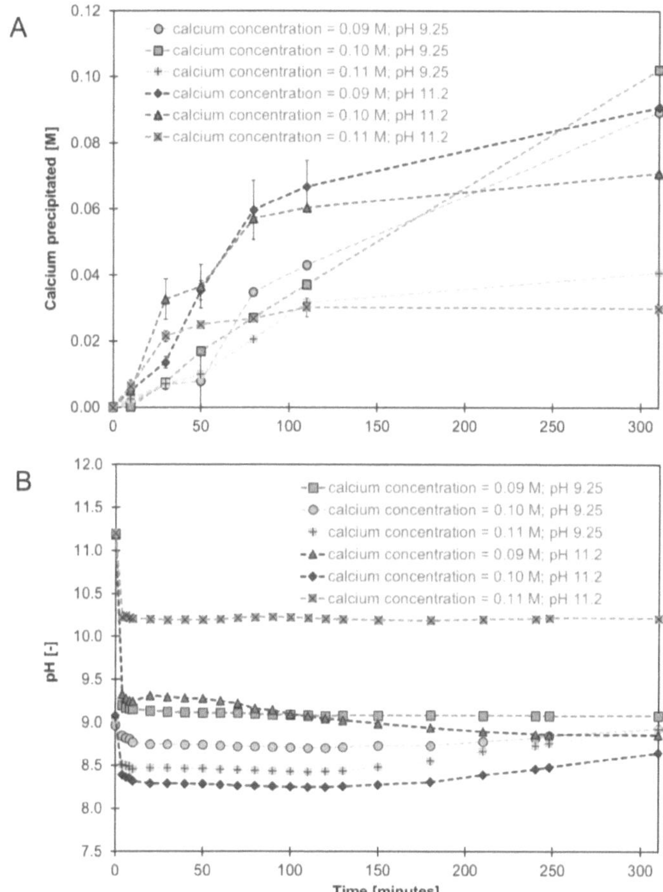

Figure 4-16: The calcium precipitated (A) and pH (B) for cementation medias with different starting influent calcium ion concentrations and pH values.

At the end of the experiment, 100%, 100% and 37% of the initial calcium precipitated for the three solutions, 0.09 M, 0.10 M and 0.11 M respectively. The changing pH in the solutions is shown in Figure 4-16B. For solutions with the same pH of 9.25, the higher the influent calcium concentration, the more significant the drop in pH in the first five minutes of the experiment. The pH of the influent concentration of 0.09 M dropped to a pH of 9.2. The higher influent calcium concentration of 0.10 M dropped further to 8.85, and the highest initial calcium concentration of 0.11 M dropped the furthest to 8.51. The pH for both solutions with higher initial calcium concentrations rose gradually after their initial drop, and by the end of the

Figure 4-16B shows the course of the pH in each flask of solution (cementation media) with varying initial calcium concentrations and pH values. Comparing solutions with the same influent calcium concentration but different initial pH values showed several variations. For an initial influent calcium concentration of 0.1 M, the solution with the initial pH of 9.25 precipitated more calcium carbonate by the end of the experiment and effectively used all the calcium for precipitation. The solution with the higher initial pH (11.2) did not precipitate the entirety of the initial calcium concentration but only 71% of it. Additionally, the rate of calcium carbonate precipitation for the solution with the highest pH was greater initially, whereas the solution with the pH of 9.25 had a relatively constant rate of calcium carbonate precipitation and eventually surpassed the rate of the solution with the highest pH (11.2).

For the highest initial calcium concentration of 0.11 M, MICP in both the pH solutions were affected negatively. However, the higher pH was affected more negatively, and by the end of the experiment, only 37% and 27% of the initial calcium concentration had been precipitated respectively. The percentages were less than those found for the lower initial calcium concentration, which ranged from 71% to 100%. Similarly, the initial rate of precipitation for the solution with the higher initial pH was greater, however after the rate decreased, the solution with the lower initial pH rate surpassed it, precipitating a more considerable amount (37%).

The effects of the initial pH and calcium concentrations on the viability of the bacteria cells were monitored by assessing urea hydrolysis in the solutions. The samples were tightly sealed in Schott bottles, preventing any escape of ammonia gas that might have formed. The ammonium concentration was measured before and after the duration of the experiment, as seen in Figure 4-17. At the end of the experiment, the solution with a pH and calcium concentration of 9.25 and 0.10 M respectively, hydrolysed the most urea as the ammonium ion concentration was the highest (0.16 M). This correlated with the maximum amount of calcium carbonate precipitated (Figure 4-16). However, its counterpart with the same initial calcium concentration (0.10 M) but at a higher pH (11.2), hydrolysed significantly less urea with an end ammonium concentration of 0.075 M. Similarly, the amount of calcium precipitated for this solution (0.1 M, pH 11.2) was significantly less than its counterpart (0.1 M, pH 9.25) (Figure 4-16). Secondly, both solutions with an initial calcium influent concentration of 0.09 M hydrolysed roughly the same amount of urea at 0.121 M and 0.125 M for a pH of 9.25 and 11.2, respectively. The solutions with the highest initial calcium influent concentration (0.11 M) hydrolysed a small amount of urea leaving an ammonium concentration of 0.42 M and 0.31 M for a pH of 9.25 and 11.2, respectively. This corresponded to low values of calcium carbonate precipitation (Figure 4-16).

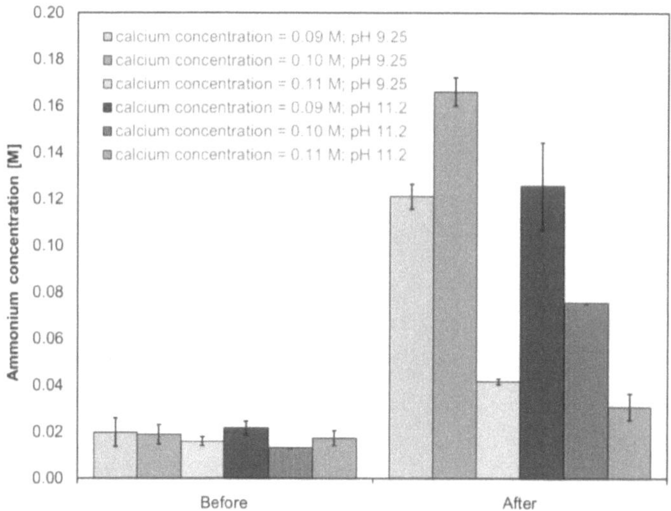

Figure 4-17: The ammonium ion concentrations of solutions with varying initial calcium concentrations and pH values before and after the duration of the experiment.

4.5 Storage of Cementation Media

Figure 4-18 shows the ammonium and calcium ion concentrations for two different cementation medias, prepared according to section 3.4, remained constant for the first four days. However, on the sixth day, the concentration of both the ammonium and calcium ions deviate for the cementation media with a pH of 9.25. This deviation accelerates from the 4th to the 8th day where the calcium concentration decreased from 0.023 M to 0.016 M to 0.003 M, and the ammonium concentration increased from 0.040 M to 0.047 M to 0.085 M. Similarly, on the eighth day the cementation media with a pH of 11.2 began to deviate and the ammonium concentration increased from 0.040 M to 0.048 M and the calcium concentration decreased from 0.023 M to 0.016 M. The derived data for this experiment can be found in Appendix E (Table E-4).

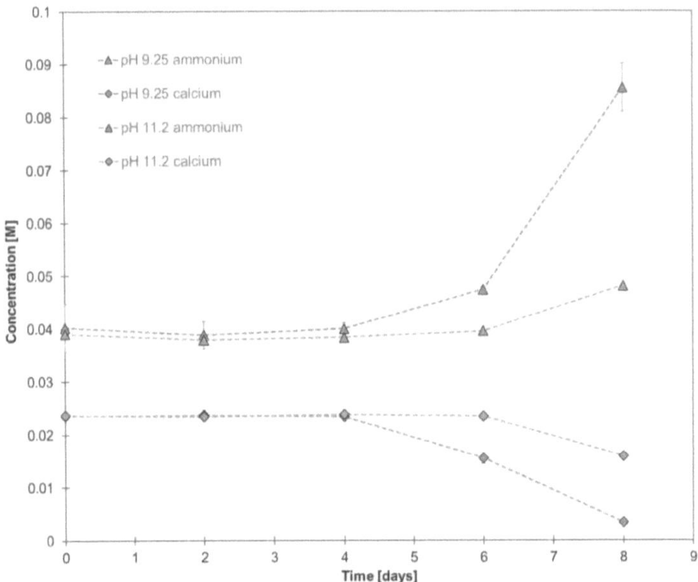

Figure 4-18: The calcium and ammonium ion concentration during storage of stabilised urine, with no excess calcium chloride, for two different initial pH values.

4.6 Alternative Nutrient Media

4.6.1 Physio-chemical Characteristics

Table 4-2 shows the various physio-chemical characteristics of the alternative nutrient medias (ANMs). Brewhouse yeast (BW) contained the highest concentration, by a considerable amount, of TKN, Org N, Org P and TP compared to LML and whey. The nitrogen (TKN) content of BW was 11 296 mg/L and is greater than the nitrogen content of ATTC®1376 which measured 3 900 mg/L and LML and whey measured less than ATTC®1376. BW's starting pH was much lower than the pH of 9 used to grow ATTC®1376 in the labs. The LML media had a pH of 6.93, which was closest to that found in the ATTC®1376 media (pH). Both the whey and BW had significantly higher concentrations of organics (COD) than LML. The theoretical COD in ATTC®1376 media was calculated to be 73 000 mg/L, which is greater than any of the alternative medias.

Table 4-2: shows the concentration of the physio-chemical characteristics: Total Keigedhal Nitrogen (TKN), Organic Nitrogen (Org-N), Total Phosphorous (TP), Organic Phosphorous (Org-P) and Chemical Oxygen Demand (COD) for each of the identified ANM.

Alternate nutrient media	Concentration [mg/L]		
	Brewhouse yeast (BW)	Lactose Mother Liquor (LML)	Whey
TKN (mg/L)	11300	1090	2029
Org N (mg/L)	2560	29	580
TP (mg/L)	690	235	112
COD (mg/L)	17470	4820	43770
OrgP (mg/L)	256	142	35.2
OD (-)_	0.699	0.197	0.54
pH (-)	4.81	6.93	6.24

4.6.2 Growth Testing

Figure 4-19 shows the change in cell density of the different ANMs from the experiments described in section 3.9.1. The full data set can be found in Appendix F. Each growth curve shows the OD value over time for the bacteria grown in each of the ANMs. The graph can be described by the phases indicative of bacterial growth (lag, exponential, stationary). At the beginning, the gradient shows a slow rate of growth while the bacteria adapt to their growing conditions and prepare for growth; this is indicative of a lag phase. The lag phase for ATTC®1376, LML and whey occurred during the first 2 hours. However, the BW's lag phase was longer than that of its counterparts and remained at a slow growth rate for four hours.

After the lag phase, the gradient of the bacteria grown in all four medias were at their steepest as they grew and adapted to their environment faster (exponential phase). Both the LML and ATTC®1376 media reached their maximum ODs at 14 hours; the peak OD was greatest in the ATTC®1376 media (1.15) than in the LML (0.97). After which it levelled off in the stationary phase and experienced a slow decline. After the first 6 hour of reaching their maximum, the OD in the ATTC® 1376 media dropped by 0.020, while the LML dropped by only 0.134. Conversely, the growth rates in the BW and whey after the lag phase were decidedly less steep and met their maximums at 20 hours (6 hours after both LML and ATTC®1376) reaching OD's of 0.67 and 0.51 respectively.

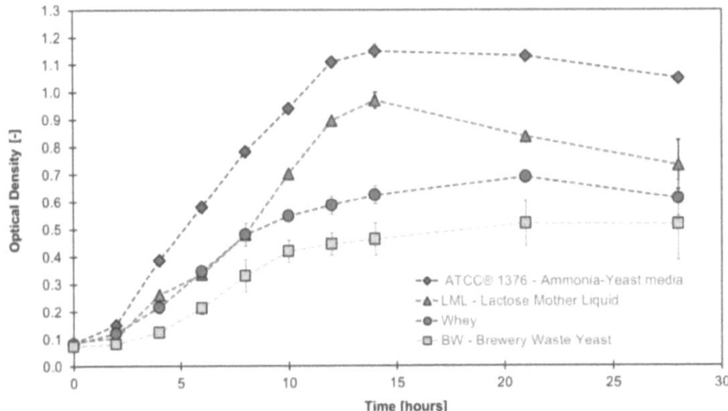

Figure 4-19: The bacteria concentration (optical density) grown in four different nutrient medias: ATTC®1376, Lactose Mother Liquor (LML), whey, Brewhouse yeast (BW).

Figure 4-20 shows images taken of bacteria grown in different nutrient medias under the microscope. Figure 4-20A shows the rod-shaped *S. pasteurii*, 2-2.5 µm in length, grown in the ATCC® 1376. The bacteria grown in the LML culture (Figure 4-20B) had the same sized rod-shaped appearance, as seen in Figure 4-20A. When observing the bacteria grown in the whey under the microscope Figure 4-20C, the most prominent bacteria is an angular shape, about 5 µm in length. However, rod-shaped bacillus bacteria mirroring the ones found in the ATTC®1376 were observed in a smaller concentration. The bacteria grown in the BW shown in Figure 4-20D also mirrored that found in the ATTC®1376 media but were spaced more sparsely.

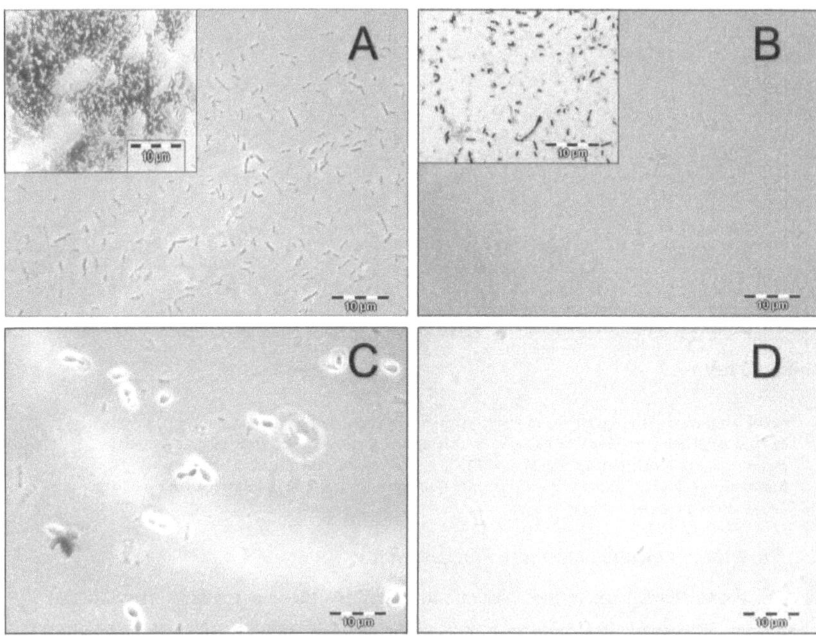

Figure 4-20: Bacteria under the microscope after being grown in different nutrient mediums overnight: (A) ATTC®1376, (B) Lactose Mother Liquor (LML), (B) whey, (D) Brewhouse yeast (BW).

Figure 4-21 shows the CUA plates before being plated with the bacteria (left) and 48 hours after being streaked with bacteria from the various nutrient medias at the end of the experimental run (right). Each plate corresponds to a different bacteria culture. Colour change in the plates indicated a pH change by turning red or pink as a result of urea degradation. Minimal to no colour change was observed in the plate streaked with BW, indicating that the bacteria viability was strongly affected by the growth medium. The colour change on the whey plate was not significantly greater than BW's, and when examined, there appeared to be another type of bacteria growing on its surface. Bacteria culture grown in the ATCC® 1376 and the LML contained viable cells as a complete colour change was observed, showing the diffusion of products due to hydrolysis. A quantified variance in ureolytic activity between the two medias cannot be determined as the CUA test only serves to indicate ureolytic activity.

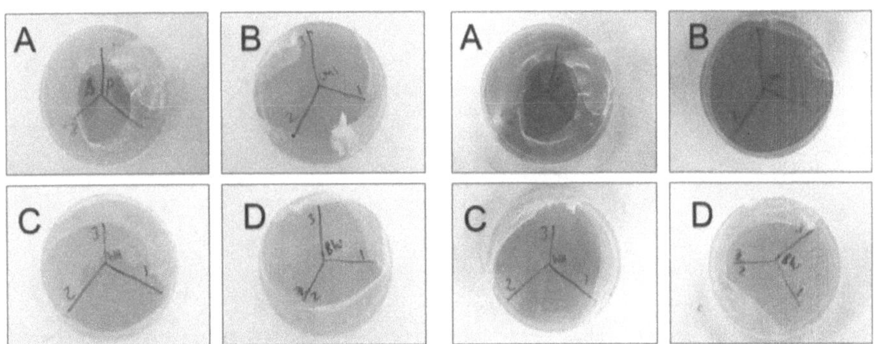

Time = 0 hours   **Time = 48 hours**

Figure 4-21: shows 1) CUA plates before being streaked with bacteria (T = 0 hours) and 2) CUA plates 48 hours after being streaked with bacteria. Each of the four plates corresponds to the four different medias used to grow the bacteria A) ATTC®1376, B) Lactose Mother Liquor (LML), C) whey, D) Brewhouse yeast (BW). Each plate is divided into three sections and streaked with the same media from each of the three triplicate flasks.

4.6.3 The Rate of Calcium Carbonate Precipitation

Figure 4-22 shows the change in the calcium concentrations for four solutions of synthetic stabilised urine after being inoculated with the varying bacteria cultures, cultivated using the three (ATTC®1376, LML, whey, BW). The calcium concentration in the three ANMs dropped more rapidly than that of the ATTC®1376 media in the first 2 hours to an average of 0.78 g/L but increased again until the 6th hour to an average 0.87 g/L. After six hours t,he calcium concentrations of the three ANMs began to decrease again. At the end of the 32 hours, the calcium concentration in the LML solutions decreased to a similar value as that of the ATTC®1376 solutions, both having almost depleted all the calcium ions within the liquid reflecting the final concentration of 0.022 g/L and 0.028 g/L respectively. The whey and BW's calcium concentration have both decreased but at a far slower rate, depleting only half of the initial calcium ion concentration within the synthetic urine solution reflecting a final concentration of 0.542 g/L and 0.503 g/L respectively.

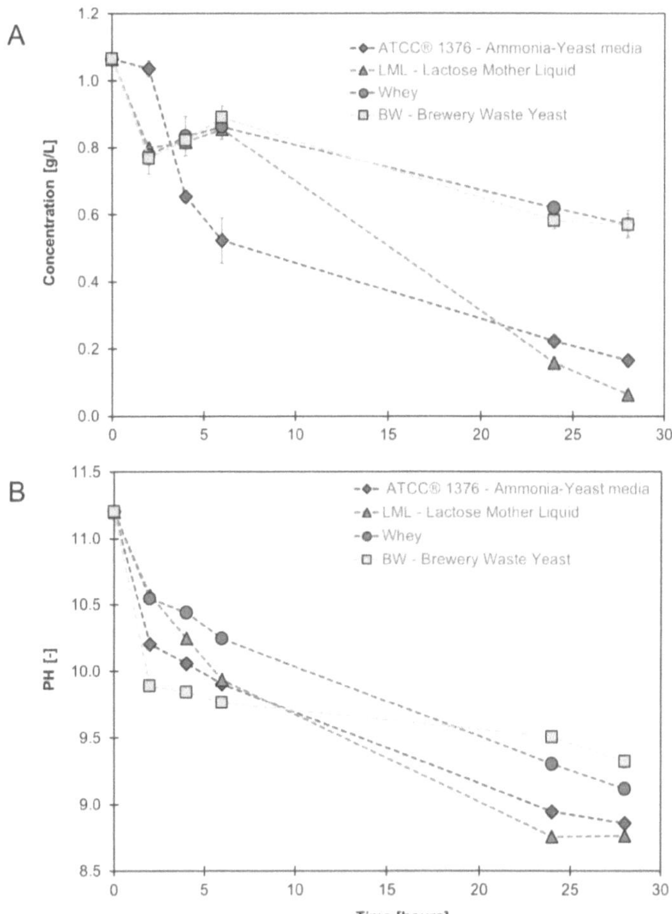

Figure 4-22: The calcium ion concentration (A) and the pH (B) of cementation medias inoculated with concentrated bacteria cultures grown in various different growth mediums (ATTC®1376, Lactose Mother Liquor (LML), whey, Brewhouse yeast (BW).

The pH of the BW decreased more rapidly than its counterparts and tapered off at a pH of 9.8, decreasing at a slower rate after that. The ATTC® 1376 and LML decreased at similar rates, and at 24 hours the gradient of both pHs had flattened out until the end of the 28-hour period, finishing with a pH of 8.8. The whey solution's pH decreases at a slower rate than the BW, ATTC® 1376 and LML, finishing with a pH of 9.0.

During the experiments, colloid precipitates formed in all the solutions. However, the ATTC® 1376 and LML solutions had signs of scaling on the bottom. After the 28 hours, the volume of precipitated calcium carbonate was measured and is displayed in Figure 4-23. Additionally, the theoretical amount calculated is also displayed in Figure 4-23. Both the ATTC® 1376 and LML solutions produced the most calcium carbonate precipitate and whey, and BW produced significantly less. The theoretical masses are on average 161 mg more than the measured masses, but still, follow the trend that depicts ATTC® 1376 and LML having far greater precipitation than both the whey and BW.

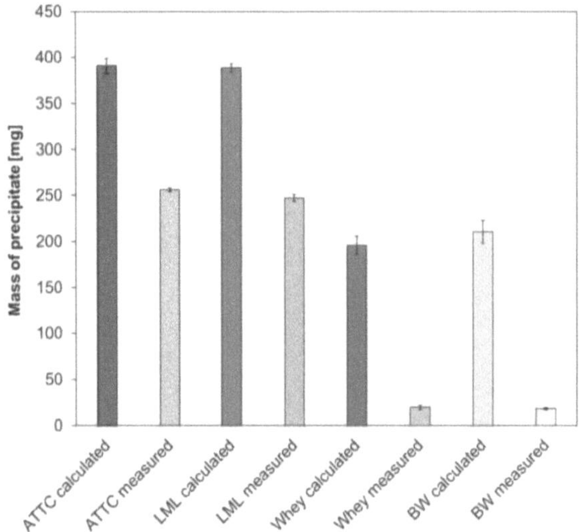

Figure 4-23: A bar graph showing the actual measured mass of precipitates formed in the varying solution during the experiment as well as the theoretically calculated mass of precipitates formed.

Chapter 5 Discussion

This chapter discusses the findings observed in Chapter 4 as well as its alignment to existing research. It specifically deals with objectives 2-5 of this book outlined in Chapter 1 which pertain to: testing the possibility of producing a bio-brick from human urine using the MICP process; improving the system design to allow the optimal influent calcium concentration, retention time and the number of treatments in relation to the resultant compressive strength; establishing a relationship between treatment duration and bio-brick strength; understanding how the ionic strength and the pH of the urine being fed into the bio-brick mould, affects the MICP process reducing the cost of the process by investigating industrial waste streams as a viable, alternative food source for the bacteria governing MICP. The chapter is structured to follow the order in which the objectives are mentioned. Included in the chapter is a discussion on the experiment conducted on eliminating smell via after treatment washes and the advantages and disadvantages of using a cementation media with a lower starting pH.

5.1 Feasibility of 'Growing' a Bio-brick

Figure 4-1 shows the initial steps of the MICP process on a microscopic scale. Figure 4-1 shows both the *S. pasteurii* colonised with the loose sand and Greywacke aggregate mixture as well as the calcium carbonate (calcium carbonate), which cements the loose sand particles together. Over time the calcium carbonate formation results in the formation of a solid in the shape of the mould – in this case, it was the shape a brick. The X-Ray Diffraction (XRD) analysis further substantiated that the solids formed during these experiments were calcium carbonate in the crystal structure of vaterite. This is consistent with the results reported by other researchers (Choi et al. 2017). Schematic 4-2 shows the process of MICP on a macroscopic scale whereby a single grain of Greywacke aggregate before and after the MICP process is shown. From the XRD analysis, it can also be established that the sand grains and aggregate mix are being cemented together with calcium carbonate. Finally, the solid bio-brick produced in experiment B2 showed that a solid in the shape of a conventional brick could be produced using the MICP process and the urea present in human urine.

5.2 Effects of Varying Starting Influent Calcium Concentrations

Experiments A1, A2 and A3 showed that for influent calcium concentrations of 0.045 M and 0.09 M, all the calcium fed into the bio-brick moulds was utilised to produce calcium carbonate. However, when the influent calcium concentration was raised to 0.11 M, more than 75% of the calcium left the moulds in the effluent stream, reflecting the point where the influent calcium concentration begins to affect the microbial activity negatively. Both bio-brick experiments. It was thus decided that the highest achievable starting influent calcium the bio-bricks system could withstand was a concentration of 0.09 M. However, this only accounts for the concentrations tested during the various experiment, so the actual maximum achievable starting influent calcium concentration could lie between a concentration of 0.09 M and 0.11 M. In addition, in the subsequent experiment R1, a new influent calcium concentration of 0.095 M was tested, and by the 6th treatment the calcium efficiency was 99%, and as a result, it became the newly established maximum. This further showed that the highest tested starting influent calcium concentration the bio-brick system could operate at was 0.95 M.

The results also showed that the influent calcium concentration was the limiting factor as the urea usage efficiency was consistently lower than the calcium usage efficiency. In order to increase the urea usage efficiency, the influent calcium concentration needed to be increased. However, a higher influent calcium concentration negatively affected the MICP process, thus limiting the amount of calcium carbonate that was able to form.

5.3   Effects of Stepwise Influent Calcium Concentrations

To improve the efficiency of the bio-brick production process, more of the urea present in the cementation media should be utilised. Adding calcium to the cementation media to achieve an equimolar amount with the existing urea would maximise the potential calcium carbonate precipitated. However, the bacteria cannot survive in an environment with high ionic strength (Hammes & Verstraete, 2002). Therefore, the critical parameter considered for optimising the MICP process was the influent calcium concentration to see if the bacteria could survive gradual increases in the ionic strength of the solution as a result of calcium chloride addition. From experiments A1, A2 and A3 an initial calcium concentration of 0.11 M resulted in low calcium usage efficiencies, which improved over time. Experiments C1 and C2 investigated if a stepwise increase during the experiment would improve the efficiency of the process. The stepwise increase in C1 showed that a point was reached at 0.13 M when urea hydrolysis dropped dramatically, reflecting an adverse effect on the microbial community. It was established that 0.11 M was the optimum tested influent calcium concentration as the maximum value before this concentration was where adverse effects were experienced. When the influent calcium concentration remained at 0.13 M, urea hydrolysis continued to decrease on a downward trend over the three days after the concentration had been increased. It was assumed that, if the influent calcium concentration was not decreased, the bacteria would continue to show decreasing signs of ureolytic behaviour until eventually no ureolytic activity was observed, resulting in a significant drop in the calcium usage efficiency. In addition, the experiments showed that the calcium usage efficiency and urea hydrolysed can be increased to higher values if the influent calcium concentration is reduced. This was shown by an increase in urea hydrolysis

Experiment C2 built on the findings from experiment C1 by increasing the influent calcium concentration from 0.11 M by smaller increments. An indication of a point where the influent calcium concentration begins to affect the microbial activity negatively was observed at the beginning of a treatment with 0.12 M, but the effluent reflected a recovery after 4 treatments, indicating that the microbial activity could be revived, and the bacteria could adapt. This revealed that adopting small increment increases in the influent calcium concentration allowed for the maximum influent calcium fed into the system to be 0.12 M. The maximum influent calcium concentration found in this study was lower than the optimal calcium concentration of 0.22 M found in other studies that only used urea and calcium chloride (Okwadha & Li, 2010).

The pH in the effluent ranged between 8 and 9.4 in experiments A1, A2, C1 and C2. It was observed that as the influent calcium concentration increased, the effluent pH decreased. These values reflected a similar range of pH measurements observed by Henze & Randall (2018) . In addition, research shows that most calcium carbonate precipitation occurs under alkaline conditions between a pH of 8.7 and 9.5 (Anbu et al., 2016). A pH of 9.25 has been identified as the optimal pH for urea hydrolysis (Lauchnor et al., 2015). An increase in calcium carbonate precipitated shows a shift in the effluent pH, the more calcium in the influent, the lower the pH of the effluent. The highest calcium influent concentrations decreased the pH to below 8.7, as a result of the expulsion of H^+ ions from the bacteria cells due to a calcium influx (Hammes & Verstraete, 2002). Precipitation also results in a pH decrease. A high ionic strength causes a calcium influx, which disrupts the cells physiological processes, negatively impacting the microbial community (Kistiakowsky & Shaw, 1953). In both experiments, C1 and C2, a pH lower than 8.6 corresponded to a decrease in the amount of urea hydrolysed. This indicated a decline in the bacteria's ureolytic activity as a result of the calcium influent concentration values of 0.12 M and 0.13 M being fed into the system. In addition, a pH of above 10 shows negligible urea hydrolysis, signifying that MICP had stopped working (Henze & Randall, 2018). The bacteria's ability to shift the pH to where it would facilitate urea hydrolysis had halted. This phenomenon was observed in experiment A3.

5.4 Retention Time

The calcium and urea efficiency were generally unaffected by retention time, except for a retention time of 1 hour, which saw a decrease in both calcium and urea efficiency. Decreasing the retention time to 1 hour caused the pH to increase to above 9.3. This indicated that the retention time was too short to sufficiently allow for the bacteria to adapt to their new environment. Compared to the other retention times tested, 1 hour had decreased values for urea hydrolysis (0.03 M less), urea usage efficiency (4% less) and calcium usage efficiency (14% less). These values were trending downward, which indicates that the bacteria were negatively be unsustainable with a greater number of treatments. It was hypothesised that increasing the retention time above 4 hours could result in decreased ureolytic activity due to the bacteria being deprived of its food source. However, the 8-hour retention time only saw a slight decrease in precipitated calcium. A smaller retention time that maintains high usage efficiencies is desired, and as a result, a retention time of 2 hours should be used for further MICP processing when using urine as a source of urea.

5.5 The Relationship Between Compressive Strength and Number of Treatments

The compressive strength of the bio-bricks is dependent on the amount of calcium carbonate deposited in the bio-brick moulds, and its effectiveness in cementing loose sand particles together. The results from section 4.1.4 show that the amount of calcium carbonate produced is dependent on the number of treatments, the greater the number of treatments, the higher the compressive strength. The compressive strength also increased when the influent calcium concentration was increased in a stepwise manner. The highest compressive strength achieved was 2.7 MPa, greater than compressive strengths achieved by bio-columns grown from synthetic urine (0.9 MPa) (Henze & Randall, 2018), and greater than MICP studies done by (Qabany & Soga, 2013) (0.35 – 1.3 MPa) and (Choi et al., 2017) (0.88 – 1.1 MPa). The strength surpassed the standard strength of 40% limestone bricks measuring 0.75 MPa but was weaker than conventional face and non-face bricks measuring 9 to 12.5 MPa and 3 to 10.5 MPa respectively (SANS 227, 2007).

The considerable variation in the compressive strengths during some of the experiments could be attributed to the different bacteria colony sizes, liquid flow patterns and the surface area of the sand (Choi et al., 2017). The relationship between the compressive strengths and the number of treatments used to produce a bio-brick (with the same influent conditions) was found to have a positive, linear relationship.

5.6 Washing

The washing experiment showed that washing the bio-bricks could help remove the strong ammonia smell released after opening the bio-brick moulds. However, for one bio-brick, this would require around 2.5 L of water, which contradicts the projects efforts to reduce water consumption in this project. It can be noted that a large portion of the smell dissipates after 3-5 days of the bio-brick being left in the open with ventilation. Economically and environmentally, drying the bio-bricks would be a better method to remove the ammonia smell.

5.7 Ionic strength, Influent PH and Storage

The effect of ionic strength was tested using the change in ammonium concentration as a representation of the amount of urea degraded. The results showed that at higher ionic strengths, enzymatic urea hydrolysis was significantly reduced. This could be attributed to the bacteria not surviving in environments with extremely high ionic strengths. Additionally, the ionic strength (0.59 M) corresponding to maximum influent calcium concentration (0.12 M) the bio-brick system could withstand without adverse effects on the microbial community is similar to the ionic strength found in work by Whiffin, Van Paassen & Harkes (2007). The ionic strength for the peak $CaCO_3$ precipitation found in Whiffin, Van Paassen & Harkes (2007) was equivalent to 0.6 M, which corresponded to a cementation media with a calcium ion concentration of 0.15 M.

To improve the efficiency of the MICP process, it is critical to maintain a low ionic strength. This presents a challenge when using urine as its ionic strength is initially high due to the varying ions present in it. A high ionic strength could be avoided by removing the unwanted ions by first recovering urea from the urine after stabilisation. The urea could then be mixed later with calcium chloride. However, this would mean utilising unnecessary amounts of water and energy and increasing the overall cost of the process. Alternatively, a mutant strain of bacteria could be developed that can survive in environments with high ionic strengths. Previously researchers have experimented with improving MICP by means of creating an adapted strain of *S. pasteurii* (Achal et al., 2009). Urease-producing halophiles that had a maximum activity in high salt environments were discovered in research by Mizuki et al. (2004). By choosing a naturally occurring urease producing bacteria that behaves in a similar way, with the ability to survive hypersaline environments, could be another solution to the issue of ionic strength.

The results of the pH experiment from section 4.4 indicate that the influent pH of the cementation media influences MICP. For the cementation media's initial pH used in the bio-brick moulds (11.2), the pH shifts quickly as the bacteria is added, to a pH where MICP can occur (~9) for high initial calcium concentrations up to 0.1 M. However, at the highest initial calcium concentration tested (0.11 M) the pH fails to shift significantly, which is reflected in a reduction in calcium carbonate precipitated and urea hydrolysed. This inhibitory effect can be attributed to the higher ionic strength associated with the increased calcium concentration. However, the results show duality in the effect of both the ionic strength (initial calcium concentration) and the initial pH. Using an initial pH indicative of the optimal pH for MICP (9.25) allowed for a higher initial calcium concentration to have a less inhibitory effect on MICP. The bacteria had the ability to shift the pH for higher initial calcium concentrations at an initial pH of 9.25 where it did not for an initial pH of 11.2, which resulted in an increased amount of calcium carbonate precipitated and urea hydrolysed in the solution with the initial pH of 9.25. This is analogous to the solution with a lower starting pH, having an increased ability of enzymes to catalyse at higher ionic strengths, which aligns with research done by Chaplin & Bucke (2017). This is possibly due to the ionic strength having less effect on the overall charge of the enzyme molecule at a neutral pH of 9.2 (Chaplin & Bucke, 2017), likely exerting a smaller influence on the enzyme's isoelectric point. However, the degree to which the initial pH decreases the inhibitory effect of ionic strength is not substantial showing a buffer of about 0.01 M before being subjected to the negative effect of high ionic strengths.

Randall et al. (2016) have previously shown that when dosed with excess calcium hydroxide, fresh urine experiences a limited loss of urea for an extended amount of time (27 days). In this current study, the pH of the filtered liquid, after being dosed with calcium hydroxide, was lowered to 11.2 to ensure the bacteria could survive (Henze & Randall, 2018). To understand at what point urea degrades during storage is vital, as it gives an indication of the effects on the amount of urea available for MICP. Experiments assessing the length of time in which urea-rich urine, with varying pHs, could be stored, were conducted. For the first few days of operation, neither solutions pH 9.25 nor pH 11.2, showed signs of urea degradation, as the ammonium concentration remained constant. However, the experiments showed that a loss of urea in the stored liquids (pH 9.25) began to occur after 4 days of storage because of the probable contamination with urease-producing bacteria. If conditions are favourable, contamination will almost always result in urea degradation at a lower pH (Udert, Larsen & Gujer, 2003). The cementation media with a higher pH can be stored for an extended period as changes in ammonium concentration (urea degradation) only occurred later at 6 days. The MICP process can run for at least 4 days before new cementation media would need to be used and to avoid significant urea degradation; a pH of 11.2 should be used. To decrease the inhibitory effect of higher ionic strength efficiency, the initial pH in the bio-brick production process could be replaced by the optimal pH of 9.25. However, this would only increase the calcium concentration fed into the system by a small amount and the system would be subject to a shorter storage periods and would need to use a more substantial amount of HCl for decreasing the pH of the feed solution. Considerations into the higher economic and life-cycle costs of the added HCl would have to be accounted for before such adjustments are made.

5.8 Alternative Nutrient Media

The nutrient media used for the growth of the urease-producing bacteria is the most significant economic barrier for the real-world application of MICP, especially when expensive laboratory grade sources are used (Rajasekar, Moy & Wilkinson, 2017). Alternatives sources of nutrient media were experimented within section 4.6. The results identified LML (lactose mother liquor) as the best alternative media. Its growth curve reached the highest OD (of the ANMs) and growth curve of LML was less than the standard laboratory reagent and reached a lower OD (ATTC® 1376), which could be as a result of the sub-inoculation of bacteria growth in ATTC® 1376 prior to the growth experiments. This allows for the bacteria to have pre-adapted to the ATTC® 1376 media, growing faster in the ATTC®1376 during the growth experiments than its counterparts (LML, whey, BW). Conversely, brewery waste yeast (BW) reached the lowest OD, and when inoculated into the cementation media, the least amount of calcium carbonate was precipitated. This demonstrates brewery waste yeast as the least suitable ANM for the growth of *S. pasteurii*. The analysis of the bacteria cultures under the microscope (Figure 4-20) show *S. pasteurii* present in all three ANM's, however, the bacteria grown in whey shares space with cells identified as yeast cells (Deacon, 2003). A possible reason for why the bacteria grow best in the LML is as a result of the pH. The optimal pH at which *S. pasteurii* grows is 9.25 (Lauchnor et al., 2015). The highest pH of all three ANMs is LML's at 6.9, as opposed to BW and whey's pH values, which are 4.8 and 6.2 respectively. The highest OD value and amount of calcium carbonate precipitated occurred in the standard reagent ATTC® 1376, which had a pH of 9. The sudden drop in pH when the BW bacteria culture was added to the synthetic urine can be attributed to the culture's low pH of 4.8. All the pH's decreased after the bacteria culture was added but more so with the ANM with the lowest pH (BW).

The change in colour, after 48 hours from yellow to solid pink, during the qualitative CUA test, shows that the bacteria grown in LML and ATTC® 1376 can both hydrolyse urea and that urease producing bacteria are present. The CUA plates for the bacteria grown in BW and whey turned slightly pink, indicating that only some urease producing bacteria had grown. However, their ability to hydrolyse urea is far less than both the ATTC® 1376 media and the LML media.

The decrease in the calcium concentrations in all 12 flasks of cementation media indicates that urease producing bacteria had effectively grown in the different ANMs. The rate at which the calcium concentration decreased, however, gives a quantitative study on the rate of ureolysis within the cementation media. The flasks inoculated with LML culture precipitated as much calcium carbonate as the flasks inoculated with ATTC® 1376 culture, indicating that they have a similar rate of ureolysis. The slower decrease in calcium concentration in the flasks inoculated with whey and BW culture indicate a slower rate of calcium carbonate precipitation and rate of ureolysis.

The measured mass of the precipitate formed differed to the theoretically calculated precipitated mass. The theoretical mass was calculated from the change in the calcium ion concentration. The measured masses all differed from their theoretical counterparts by similar values by an average of 160 mg. This discrepancy is likely as a result of some of the precipitate left as scaling on the flasks, precipitate being lost in the oven, incinerator and in the filtering process. However, the measured and theoretical values of both the LML and ATTC® 1376 are similar and are both higher than the precipitate formed by the whey and BW cultures.

These results are consistent with Achal and co-workers, who showed that LML could be used as a nutrient source for *S. pasteurii* (Achal et al., 2009b). Therefore, LML should be considered for MICP processes. However, a continuous and consistent source of the nutrient solution would be required to ensure the large-scale operation of any MICP process.

Chapter 6 Integrated System

This chapter analyses the bio-brick production proposed in this book by means of its economic feasibility, social acceptance and policy barriers. The method of producing bio-bricks implemented in this project involves the stabilisation and collection of urine from waterless nutrient recovery urinals. The integrated system was considered in its entirety when assessing the overall cost and conducting a mass balance over the bio-brick production system, right from urine collection (producing on-site fertilisers) to the effluent leaving the bio-brick. It additionally involved further fertiliser production from the nitrogen-rich effluent leaving the system. The effluent provides an additional opportunity to recover nutrients in the form of nitrogen fertilisers, which can possibly add to the revenue of the proposed process. This is considered within the integrated system for the purpose of the costing analysis but was not included in the experimental testing of this study. The mass balance was calculated for the purpose of measuring the integrated system as a whole, to understand the quantities of the inputs and outputs, in order to calculate the raw material input costs and final products sales as accurately as possible.

Additionally, the discussion talks about the feasibility of implementing the integrated system in terms of urine collection, financial viability, social willingness and possible policy barriers. It further discusses possible paths for cost reduction, further potential revenue streams and indirect benefits.

6.1 Integrated System and Mass Balance

Figure 6-1 details the integrated recovery process and its mass balance. The novel, on-site urinals are used to collect and stabilise urine in Stage 1. During the stabilisation process, enabled by a prior dosage of calcium hydroxide, calcium phosphate is produced as a fertiliser. Scaling up the system would require retrofitting waterless recovery urinals into old systems or installing an entirely new system of urinals (Flanagan & Randall, 2018).

The liquid supernatant leaving Stage 1 has a pH of 12.5. Enzymatic and chemical urea hydrolysis is inhibited by the prior dosage of calcium hydroxide to the urinal, keeping the nitrogen in the form of urea by increasing the pH. In Stage 2, the liquid supernatant is filtered, the nutrient broth is added, and the filtered liquid is dosed with HCl to lower the pH to 11.2 ensuring bacterial survival and MICP to take place (Henze & Randall, 2018). Finally, calcium chloride is added to the liquid to increase the calcium concentration. This resultant liquid is called the cementation media and is pumped through the bio-brick mould. Before the cementation media is pumped through the system, the bio-brick mould is filled with masonry sand and an aggregate mix that is inoculated with a bacteria culture of S. pasteurii. MICP takes place within the bio-brick moulds at ambient temperature, and after a number of treatments, Stage 2 produces solidified bio-bricks for industry use.

Stage 2's effluent stream is still nutrient rich and has a high concentration of ammonium ions. This serves as an additional nutrient recovery stream where another alternative fertiliser, ammonium sulphate, can be recovered. The ammonia produced during the MICP process is considered a significant setback in the process (Choi et al., 2017). However, this study proposes a solution that converts the ammonium-rich effluent into ammonium sulphate, a widely used fertiliser (Pradhan, Mikola & Vahala, 2017). The effluent leaving Stage 3 holds the potential for potassium to be recovered as a third, alternative fertiliser. However, to date, no experimental research pertaining to the recovery of potassium from an effluent stream like the one leaving Stage 3 has been investigated. Potassium recovery was therefore, not included in the economic analysis.

The mass balance shown in Figure 6-1 was calculated for the system using the values for experiment C1, as it obtained the highest strength bio-brick. To acquire a compressive strength of 2.7 MPa, the bio-bricks were treated for a total of 48 treatment cycles over 4 days. Additionally, the masses were calculated for a daily production of 1000 bio-bricks. A large mass of urine (328 tonnes) is required for a production of 1000 bio-bricks. However, only 1% of the influent urine is actually utilised in the MICP process (~350kg). The stabilisation process produces calcium phosphate, which has proven to be an effective fertiliser (Meyer et al., 2018). The amount of calcium hydroxide fed into the system was equivalent to 10 g/L (Randall et al., 2016) which amounts to 312 kg of calcium hydroxide for a daily production of 1000 bio-bricks. The amount of recovered phosphate fertiliser produced in Stage 1 was adapted from work done by Flanagan & Randall (2018), which stated that 11.23 g of calcium phosphate fertiliser is produced per kg of urine equating to a fertiliser production of 350 kg for 1000 bio-brick production.

The mass inputs for stage 2 were separated into the inputs added to the stabilised urine to make up a cementation media and the initial inoculated sand and aggregate mix required to produce a 1000 bio-bricks. The mass of calcium chloride added was calculated to raise the influent calcium ion concentration to 0.11 M from 0.029 M, the average initial calcium concentration in the stabilised urine. Additionally, 3.5 mL of 0.1 M 32% HCl was added per litre of stabilised urine to lower its pH from 12.5 to 11.2, and this amounted to a daily usage of 127 kg. Furthermore, the 93.6 kg of nutrient broth was added to the stabilised urine stream. To produce 1000 bio-bricks, 1480 kg of sand and aggregate mix inoculated with 0.935 kg of bacteria culture was required to fill the moulds. The output of stage 2 produced 1000 bio-bricks amounting to 1830 kg.

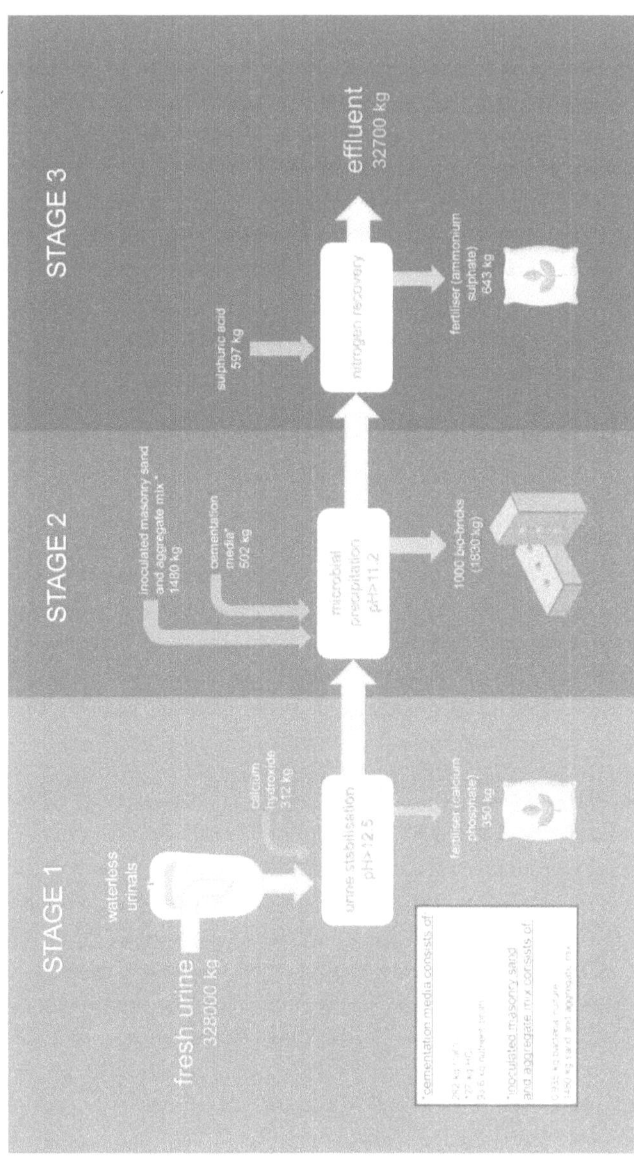

Figure 6-1: The theoretical process flow and mass balance for the integrated system for a daily production of 1000 bio-bricks. The figure shows the process in three main stages, each producing a product, calcium phosphate fertiliser, bio-bricks and ammonium sulphate fertiliser. The figure details the masses of the inputs and outputs for the entirety of the integrated system.

The MICP process converts the urea present in stabilised urine into ammonium and carbonate ions, and so the effluent leaving stage 2 is rich in ammonium, equivalent to the concentration of urea in the stabilised urine of 0.39 M (Randall, 2016). Stage 3 produces ammonium sulphate when the ammonium ions react with the added sulphuric acid, which has been shown to be an effective method for producing ammonium sulphate from urine (Pradhan, Mikola & Vahala 2017). An efficiency of 80% was assumed for this process. The effluent from stage 2 and the mass of ammonium sulphate produced was calculated using stoichiometry according to equation 6-1 below. Thus, to produce 1000 bio-bricks, 597 kg of sulphuric acid would be used to recover 643 kg of ammonium sulphate fertiliser. Lastly, an effluent equivalent to 327 tonnes leaves the integrated system after stage 3.

$$2NH_3 + H_2SO_4 \rightarrow (NH_4)_2SO_4$$

6-1

6.2   Economic Evaluation

A preliminary economic evaluation was conducted, which only included the cost of raw material inputs and the sales of goods produced. The costing was calculated according to the masses shown in Figure 6-1. Additionally, the costing was calculated based on producing the bio-bricks locally (Cape Town) and a daily output of 1000 bio-bricks. This daily output is only a small portion of the output of conventional brick kilns in South Africa. A single plant that mixes and shapes bricks via mechanisation can produce up to 700 000 bricks a day (CBA, 2016).

Table 6-1 displays the incomes and expenses considered in the economic evaluation. Firstly, it was assumed that the urine was obtained at no cost. The nutrient recovery urinals (NRUs) were dosed with calcium hydroxide at a cost of 2.87 ZAR/kg. This step recovered calcium phosphate on-site which could be sold for 1.85 ZAR/kg incurring a daily cost and sale revenue of ZAR 895.44 and ZAR 6475 for 1000 bio-bricks respectively. It was assumed that the drying of the solid fertiliser would occur at room temperature, incurring no extra cost for energy. The reduced nutrient loads on the wastewater treatment plants (WWTPs) provides an indirect, beneficial decrease in costs and fewer negative environmental impacts but these benefits were not considered in the economic evaluation. For Stage 2, the incurred cost of calcium chloride was calculated to be ZAR 1199 calculated from 4.25 ZAR/kg. Furthermore, the cost of using 127 kg of HCl was calculated using the specified price of 3.76 ZAR/L and a density of 1.16 g/m^3 amounting to R412 Additionally, the sand and aggregate mix costs 80 ZAR/tonne amounting to a total expense of ZAR 118.4.

The most substantial input cost would have been the nutrient media required for culturing the bacteria and the nutrient broth used in the cementation media. The prices of laboratory grade nutrient broth and ATTC[®] 1376 growth media was ZAR 1193 per kg (Titan Biotech, 2019) and for the cost per L of urine is ZAR 4.25 (Sigma-Aldrich, 2018). When incorporated into the costs, these made up more than 95 % of the total costs and resulted in a net loss. However, section 5.7 indicates that LML can be effectively used as an ANM. To acquire the LML, it is assumed that the waste stream can be obtained at no cost, with transportation expenses ignored. The LML can be used both as a nutrient broth and a bacteria culture and as a result, both costs are represented as zero. Transportation cost for the collection of the treated urine would contribute to the overall cost of the process but this was not considered in this economic evaluation but should be investigated further in a more detailed analysis.

The revenue incurred from the sales of the bio-brick was calculated according to the price of a conventional brick. If bio-bricks were to enter into the market, it was assumed that the bio-bricks were sold at market value as the industry are price takers rather than price setters and it was assumed that and 'eco' premium would not hold in this specific market. The price of a bio-brick was taken as ZAR 2.75 which would receive daily sales of ZAR 2750.

Table 6-1: A table showing the expenses and incomes for the inputs and outputs for the integrated urine treatment system that produces 1000 bio-bricks per day.

Item	Costs	Amount [kg/day]	Cost [ZAR/day]	Reference
Expenses			3085	
calcium hydroxide	2.87 ZAR/kg	312	895.40	(Chipako, 2019)
calcium chloride	4.25 ZAR/kg	282	1199.00	(Proteachemicals, 2018)
HCl	3.76 ZAR/L	127	411.70	(Kemcore, 2018)
nutrient broth		0.00	93.6	
bacteria culture		0.00	0.94	
sand & aggregate mix	0.08 ZAR/kg	1480	118.40	(Afrisam, 2018)
sulphuric acid	1420 ZAR/m^3	597	460.70	(Alibaba, 2019a)
Incomes			10420	
calcium phosphate	18.50 ZAR/kg	350	6475.00	(Coathers, 2016)
bio-brick	2.75 per bio-brick	1000 bio-bricks	2750.00	(Builderswarehouse, 2018)
ammonium sulphate	1.85 ZAR/kg	643	1190.00	(Alibaba, 2019b)
Total profit			**7330**	

The cost of treating the effluent from Stage 2 with sulphuric acid amounted to ZAR 461, calculated from a price of 1420 ZAR/m^3 and using a density of 1.84 g/m^3 for sulphuric acid. The revenue from the sale of ammonium sulphate fertiliser amounted to ZAR 1190. When the inputs and final product for Stage 2 are considered as a standalone process, a profit of ZAR 1020 is achieved. If Stage 2 and Stage 1 are combined the profit increases to ZAR 6600 and finally if the entirety of the integrated urine treatment process is analysed, the total profit from the preliminary economic analysis was calculated to be ZAR 7330.

6.3 Discussion

6.3.1 Mass Balance and Economics

In assessing the feasibility of producing bio-bricks, firstly, it is essential to note that a large amount of urine is required for a daily production of 1000 bio-bricks (328 tonnes of urine). Producing, 1000 bio-bricks is not an unreasonable amount for a factory to produce. However, it is far less than the amount of conventional bricks produced by industry leaders. The volume of urine required for 1000 bio-brick production amounts to approximately 23% of the Cape Town's daily urine production, as shown in Figure 6-2. Assuming that Cape Town's population is 4.2 million, the annual urine production per year amounts to 50 232 m^3 (Chipako, 2019). However, to collect 23% of the daily urine produced by Cape Town, significant adjustments and retro-fitting to our infrastructure would be required and an extensive transportation system for urine collection would need to be established, all incurring additional costs. Additionally, if all the urine produced in Cape Town was utilised, a maximum daily production of 4110 bio-bricks is possible, which is still far less than that of industry leaders. Thus, using human urine for production of bio-bricks would not be able to replace conventional brick production but could take up some of the market.

The optimisation of this system could be increased to reduce the amount of urine required for the manufacturing of bio-bricks. The limiting effect of the high ionic strength of urine results in only about 40% of the urea in urine being utilised in the MICP process. The volume of urine required for bio-brick production could be further reduced by firstly recovering a 100% of the urea from urine and then using the recovered urea for the subsequent production of bio-bricks. Recovering urea for its later use in bio-brick production would decrease the percentage of Cape Town's daily human urine required to 9%. It would also potentially reduce the urine required by an even greater degree, as the stored urea would be dissolved in water, thus negating the effect of the increased ionic effect of urine, in turn increasing the amount or urea converted into calcium carbonate. However, this is not without pitfalls as the use of water potentially eliminates the positive impact of using waterless urinals, to what degree, however, would need to be further investigated. The feasibility of collecting and utilising 9% of Cape Towns' daily urine can be assessed by comparing it to a study conducted on the collection of urine and implementation of waterless urinals based in Cape Town recently conducted by Chipako (2019). The study modelled the installation of nutrient recovery urinals in shopping malls in the Cape Town area. The system proposed collecting a total of 392 m^3 from 8 locations per year amounting to 0.8 % of the annual urine production of cape town (Chipako, 2019). The system proved feasible, economically viable and found that the transportation and logistics were not a major cost and only contributed a small fraction to the total operating costs (Chipako, 2019),

further extended from 8 shopping centres to include highly frequented areas such as large office buildings in the city centre or schools to achieve the collection required for bio-brick production.

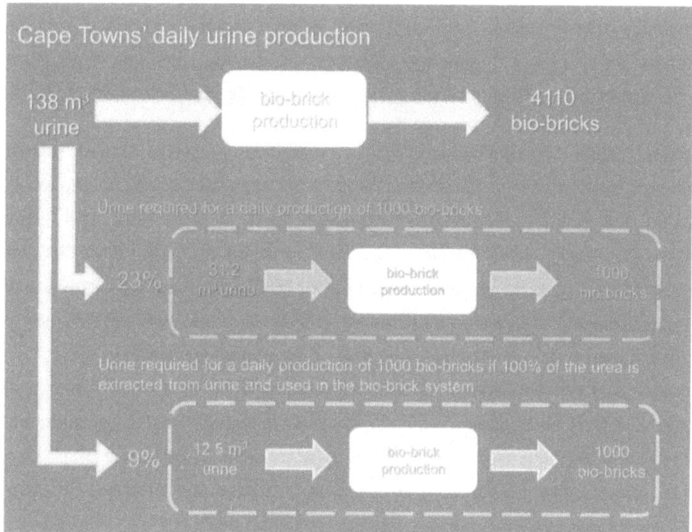

Figure 6-2: The volume of urine produced daily by Cape Town residents and the potential daily bio-brick production. The percentage of Cape Town's daily urine production required to produce 1000 bio-bricks daily, firstly if the stabilised urine stream was used directly for bio-brick production or secondly if urea was recovered from the stabilised urine to be stored and used later for bio-brick production.

Although implementing the system as is, with 40% urea utilisation, would result in using significant volumes of urine (23% of Cape Towns' population would be required to produce 1000 bio-bricks per day), only 1% of the urine is used to make one bio-brick and the remaining volume of urine could provide large volumes of fertiliser. For a 1000 bio-brick daily production process, annually, 525 tons of calcium phosphate and 965 tons of ammonium sulphate would be produced. A possible market for the fertiliser produced could be the Stellenbosch wine region, which Cape Town is situated within. The Stellenbosch wine region constitutes roughly 16% of the total vineyard surface area in South Africa, according to the South African Wine industry (SAWIS) (SAWIS, 2017). The fertiliser production from the integrated system would contribute to 53% and 31% of the annual nitrogen (763 000 kg) and phosphorous (240 000 kg) fertiliser demands for the Stellenbosch wine region respectively thus potentially further increasing the profit of the integrated urine treatment process.

This economic calculation serves as a rough indication of the financial viability of bio-brick production. Only the raw material input costs and the sales of the end products were considered. A brief analysis was conducted, as a system of this nature would require capital and operating costs that cannot be predicted by a desktop study alone since it would require key changes to our current infrastructure and WWT norms. The profit calculation for the bio-brick production as a sole entity resulted in a net positive revenue and the addition of both the calcium phosphate production and ammonium sulphate production costs and revenues to make-up the integrated system further served to increase the profit margin (ZAR 7330).

This profit could be further increased by recycling waste material like recycled concrete (Choi et al., 2017) from construction sites or limestone powder from aggregate quarries, (Choi et al., 2017) instead of the sand and Greywacke aggregate used in this study. This would simultaneously negate the impact of mining sand on the environment. BioMASON (2016) has even shown recycled glass to be an appropriate material for MICP. Additionally, by using LML as an ANM, the costs are significantly reduced by more than 95%, which achieves the 50% desired decrease in cost outlined in this dissertation's objectives. The effluent leaving stage 3 is still rich in potassium and could offer an additional recovery stream; however, further research on recovering potassium from an effluent stream such as this one would need to be conducted. Currently, the industrial methods used to recover potash (flotation, dissolution-recrystallisation, heavy-media separation, solar evaporation) are all done on mined ores or brines (Institute, 2010), and recovery of potassium in urine via k-struvite crystallization has shown to be effective. However it recovers both calcium phosphates and potassium phosphate fertilisers simultaneously (Huang et al., 2019). Both methods are unsuitable for the requirements of the integrated urine treatment system proposed in this study. Alternatively, the ammonium-rich and potassium-rich stream produced after stage 2 could also be concentrated using reverse osmosis (Thorneby, Persson & Tragardh, 1999) rather than being converted into ammonium sulphate. The research for using such a method is still underdeveloped, and a production of liquid fertiliser concentrate requires further investigation.

A detailed economic analysis was beyond the scope of this book but such an analysis is required to obtain accurate values that include capital,labour, electricity, rent and maintenance costs for the integrated process. In addition, further research and development as to the best treatment scheme is required which will further improve the economic evaluation. All the processes in the integrated system are still in the research and development stage, which makes it difficult to cost equipment accurately. However, if scaled up, it is assumed that rather than using the moulds used in this study, a mechanised spray system like that used in BioMASON production would be used (Dosier, 2016).

Ideally, urine should be treated, and fertilisers produced on-site at locations where large volumes of urine can be, for example, at office blocks, schools or universities (Etter et al., 2011).

However, the quantities required for a daily production of urine are far higher than what any on-site recovery location could produce. Therefore, other aspects would have to be factored into the business model such as transporting the urine from several commercial buildings to a decentralised resource recovery facility, where bio-bricks can be produced, and fertilisers extracted. Most solutions proposed for the recovery of nutrients from wastewater have been based on the established concepts of collection, transport and centralised systems (Ledezma et al., 2015) as conventional piped transportation and infrastructure is inadequate for transporting stabilised urine as it results in the precipitation of salts causing blockages or nutrient buildups in the pipe systems (Ledezma et al., 2015). Conversely, as stated previously, Chipako (2019) indicated that use of a 4 ton truck and one centralised resource recovery factory is financially feasible, especially in a densely populated city such as Cape Town, for the treatment of urine and its GHG emissions and energy expenditure were found to be more favourable than that of conventional wastewater (WWT). However, the study by Chipako (2019) volumes of urine significantly less than that required in this study so a decentralised system collecting the quantities required for the 1000 bio-brick daily production would have to be considered in a detailed economic and environmental impact analysis, to decide if the transportation method is feasible. Alternatively, a more efficient use of the urea present in the urine would have to be developed

The implementation of an integrated urine treatment system would offer important indirect benefits, both in terms of cost and adverse environmental effects reduction. The application of the integrated system would reduce nutrient loads on wastewater treatment plants (WWTPs), reduce water bills, reduce building maintenance costs due to fewer pipe blockages and improve environmental conditions (Chipako, 2019). Compared to the conventional production of fertilisers and bricks, the integrated system's use of energy and the impact on the environment could be significantly less (Kavvada et al., 2017). The ammonia recovery system has the potential to replace a small portion of the fertilisers produced using the energy-intensive Haber-Bosch process that relies heavily on fossil fuels, and the proposed phosphate recovery would reduce the global demand on mined phosphate rock (Chesworth, 2008). In addition, the bio-brick production provides an alternative option which could reduce our reliance on energy-intensive brick kilns.

6.3.2 Social Willingness, Product Safety Policy Barriers

The implementation of a system that recovers products for sale and uses human urine can not only be examined through an economic lens. The potential social and policy barriers should also be considered and discussed for a fuller understanding of the real-life application of this book.

It is critical to gauge the social willingness to use the waterless urinals and the products recovered from urine. Sanitation and resource recovery systems can improve the environmental sustainability of wastewater management. Yet, the social acceptability of such a new, resource-orientated sanitation practice needs to be assessed systematically (Simha et al., 2018). A survey was conducted to evaluate the willingness of university goers at the University of Cape Town to use such systems (Chipako, 2019). Of the respondents, 96% stated that they would be willing to use such technology and 79% said they were willing to eat food grown using phosphorus fertilisers recycled from human urine (Chipako, 2019).   Additionally, a recent survey at a university in the south-eastern region of the United States showed that 84% of respondents would vote in favour of urine separation in residence halls, especially when the benefit of water conservation was mentioned (Ishii & Boyer, 2016). This shows that educating users on the importance of resource recovery and the benefits of implanting such systems increases their willingness to participate. Women were less likely to want to use urinals (Lamichhane & Babcock, 2013). This could be attributed to female urinals not yet being integrated and normalised in popular social conventions. Religion also influences a user's acceptance of urine by-products as Islamic customs consider urine to be a spiritual pollutant (Drangert, 1998, Flanagan & Randall, 2018).

Additionally, a study based in India evaluated 1252 Indian consumers on their attitudes towards using products recovered from human urine (Simha et al., 2018). The attitudes were observed to be moderately positive, and 68% stated human urine should not be disposed but recycled, 55% considered it as fertiliser, but only 44% would consume food grown using it (Simha et al., 2018). Despite the fact that 65% of consumers believed using urine as crop fertiliser could pose a health risk, the majority (80%) believed it could be treated so as to pose no risk (Simha et al., 2018). The research also found a significantly stronger link between a consumer's willingness to consume urine-fertilised fruit by the price and their willingness to pay rather than their environmental attitudes (Simha et al., 2018). The study went on to state that merely appealing to people's environmental sensitivities is not sufficient for pioneering environmentally-friendly technologies like the resource recovery from urine applied in the integrated system, but that more targeted marketing messages are needed (Simha et al., 2018). Furthermore, to date, no research has been conducted on the willingness of consumers to use bio-bricks produced using urea from human urine and should be investigated further.

For the application of the integrated urine treatment system, product safety would have to be considered. Although intrinsically sterile, urine can be cross-contaminated with faeces, but less contamination is expected from urine collected directly from urinals (Flanagan & Randall, 2018). In addition, the high pH of the solution during urine stabilisation ensures that viruses and pathogens are killed. The effluent leaving the integrated system would still have traces of pharmaceuticals. Lienert and co-workers found that 64% of 212 measured pharmaceuticals were found in excreted urine (Lienert, Bürki & Escher, 2007). A challenge is presented due to 98% of the pharmaceuticals and hormones remaining in the liquid fraction after fertiliser precipitation (Ronteltap, Maurer & Gujer, 2007). This poses little threat to the quality of the solid fertiliser products or bio-bricks produced but presents a problem for the treatment of the effluent or liquid fertiliser.

In South Africa, current waste re-use and recovery are subject to waste management regulations and controls in order to regulate the full life cycle of the waste (Oelofse & Godfrey, 2008). However, internationally, the controversy surrounding the barriers to potentially recyclable 'waste' emerges from discrepancies associated with terminology specific to 'waste' regulation (Oelofse & Godfrey, 2008). Generally, in current legal definitions, the term 'waste' includes materials that are technically suitable for recovery and re-use. However, waste suitable for recovery and re-use becomes subject to the same regulations as other waste-streams (not suitable for recovery) by being under the same umbrella definition of waste (Oelofse & Godfrey, 2008). There are currently at least two legal definitions of waste in South African legislation, one in the Environment Conservation Act (ECA) (No. 73 of 1989) and the other in the National Water Act (NWA) (No. 36 of 1998). These definitions are both broad but restrictive, protection-based definition of waste ultimately becomes a barrier to industry and the successful implementation of the waste hierarchy shown in Table 6-2 (DEAT, 2006).

Table 6-2: The waste hierarchy which establishes preferred program priorities based on sustainability (Oelofse & Godfrey, 2008).

Cleaner production	Prevention
Cleaner production	Minimisation
Recycling	Re-use
Recycling	Recovery/Reclamation
Recycling	Composting
Treatment	Physical
Treatment	Chemical
Treatment	Biological
Disposal	Landfill

South Africa's present approach regarding waste management and waste re-use is generally determined by the current legal definition of waste and the associated legal requirements, which is depicted in Figure 6-3. Thus, waste re-use and recycling facilities in South Africa are subjected to similar controls as waste disposal sites such as the need for an operating permit. Conversely, exemptions to permits are offered by the ECA, however obtaining this authorization can often be a lengthy, bureaucratic process. Additionally, legislation is not clear whether the reprocessed material should continue to be deemed a waste or a product.

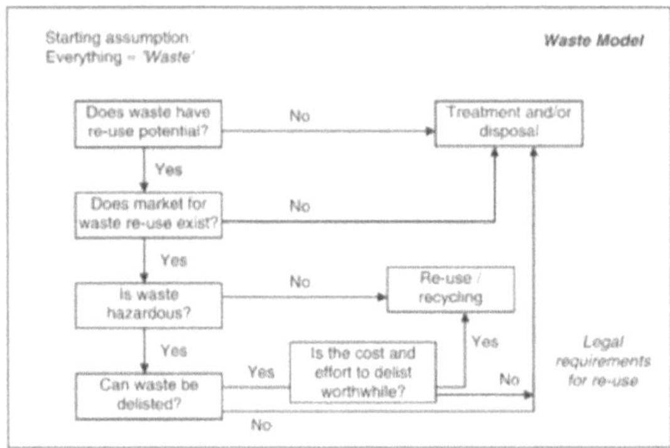

Figure 6-3: A schematic for managing waste through the waste model (Oelofse & Godfrey, 2008)

The Minimum Requirements for Waste Disposal (DWAF, 1998, DWAF, 2005) applies the precautionary principle which assumes that a waste is highly hazardous and toxic until proven otherwise (Oelofse & Godfrey, 2008). The burden of proof as to whether a waste is non-hazardous lies with the producer of the waste in question, which impedes exchanging material for re-use (Oelofse & Godfrey, 2008). This barrier would not only be encountered for recovery and re-use of urine but also when acquiring the LML waste from the dairy industry for its' re-use as most industry wastes are considered hazardous purely because of their large output volumes. Hazardous waste must be reclassified and de-listed via the process outlined in the 'Minimum Requirements for the Handling, Classification and Disposal of Hazardous Waste' (DWAF, 1998, DWAF, 2005) to prove it is not harmful to either the environment or human health. This de-listing process requires expensive specialized tests to be conducted on both the waste and the final products (Oelofse & Godfrey, 2008). Furthermore, once reclassified, the material is still defined as a waste and must uphold the regulatory requirements of its' new waste classification.

It can be assumed that urine would be categorized under hazardous waste following the minimal requirement for waste disposal and thus it is advised to follow a path for assessing if the cost to de-list and reclassify it for recycling is worthwhile. Additionally, once reclassified, an operating permit would still need to be acquired, or alternatively, an exemption for a permit could be pursued through the ECA (Environment Conservation Act).

Chapter 7 Conclusions

This final chapter summarises the research methodologies conducted in this book. Additionally, the section discusses how the book accomplished its research objectives and summarises the key findings while also offering recommendations for future research in this field.

7.1 Research Overview

A review of the existing literature on the MICP theory and application as well as urine stabilisation methods were conducted to establish and guide the equipment and methodology utilised in this book.

Bio-brick experiments were carried out to address specific research objectives. Further side experiments were conducted to test certain factors affecting the bio-brick system. Finally, experiments were conducted to find a cheaper alternative to the growth media used for the bacteria. The financial feasibility, the socio-economic effect and the South African policies pertaining to the bio-brick process that also included fertiliser recovery were assessed.

7.2 The Accomplishment of Research Objectives

The objectives of this book were to:

1. Conduct an in-depth literature review on the kinetics and theory of MICP, the factors affecting MICP, previous research and industry application of MICP and finally urea hydrolysis and urine stabilisation methods.
2. Test the possibility of producing a bio-brick from human urine using the MICP process.
3. Improve the system design to determine optimal influent calcium concentration, retention time and number of treatments in relation to the resultant compressive strength.
4. Establish a relationship between treatment duration and bio-brick strength.
5. Understand how the ionic strength and the pH of the urine being fed into the bio-brick mould, affects the MICP process.
6. Reduce the cost of the process by investigating industrial waste streams as a viable, alternative food source for the bacteria governing MICP.
7. Conduct a preliminary economic analysis of an integrated urine treatment system and briefly discuss the social and policy barriers likely to arise if such a system is implemented.

7.3 Research Conclusions

The key finding and conclusion of this book is that using the urea present in human urine can be used to grow bio-bricks using microbially induced carbonate precipitation (MICP).

Furthermore, the highest starting influent calcium concentration was found to be 0.09 M, as no adverse effects on the microbial community were observed. Additionally, in terms of a stepwise increase in the influent calcium concentration during the treatment cycle, the concentration could be raised to 0.12 M in a stepwise fashion without the microbial community experiencing adverse effects. The minimum retention time required for producing a bio-brick was 2 hours. The highest compressive strength of a bio-brick was found to be 2.7 MPa and to produce this strength required about 31.2 L of stabilised urine. A relationship between the number of treatments and the compressive strength was established, showing an increase in the compressive strength as the number of treatments increased.

Additionally, it was identified that the ionic strength of the urine had an inhibiting factor on the ureolytic activity. As ionic strength increased, the rate of urea hydrolysis decreased, and at an ionic strength of 0.8 M, almost no urea hydrolysis occurred.

It was also concluded that LML (Lactose mother liquor) could be used as an alternative growth media for *S. pasteurii* as it had comparable growth curves and calcium carbonate precipitation rates to the laboratory grade growth media (ATTC®1356). Using laboratory grade growth media is expensive, and it was established that replacing it with LML would reduce the operating cost of the integrated system by more than 95%.

The integrated urine treatment process proposed that bio-brick production should include fertiliser production thus offering additional options for the recovery of nutrients present in urine. The integrated system composed of recovering calcium phosphate fertilisers before the production of bio-bricks and ammonium sulphate fertiliser after. The mass balance over the integrated system showed that urine from 23% of Cape Towns' population would be required to produce 1000 bio-bricks per day with the current process operating conditions. Importantly, only 1% of the 31.2 L of urine is used to make one bio-brick. The remaining volume of urine could provide 53% and 31% of the annual nitrogen and phosphorous fertiliser demands for the Stellenbosch wine region, respectively. The ionic strength in the stabilised urine was found to be a limiting factor, and as a result, only 40% of the urea content was utilised in the bio-brick production which leaves significant room for further optimisation. A net positive revenue of ZAR 7330 was established between the sale of bio-bricks and fertilisers produced and the cost of raw material inputs of the integrated urine treatment process.

Certain policy barriers need to be overcome to implement the integrated system such as reclassifying the urine for its use in an industrial process. Additionally, once reclassified, an operating permit would still need to be acquired, or alternatively, an exemption for a permit could be pursued through the ECA (Environment Conservation Act). Research showed mixed reviews on whether products from the integrated urine treatment process are likely to be socially accepted and that a combined appeal to people's environmental sensitivities and targeted marketing messages could enhance people's acceptance.